WHAT EVERYBODY OUGHT TO KNOW BEFORE MOVING A DATA CENTER

BLAINE BERGER

For information regarding bulk discounts, branding requests, or custom ideas for your firm, contact the author at blaine@e-oasis.com.

Editing: Barbara McNichol

Cover Design: Gregory B. Russell

Interior layout by Kerrie Lian, under contract with Karen Saunders & Associates.

Book Web Site: http://datacentermovinghurts.com

Published in the United States by Team One Horse.

ISBN-10: 1940631009

ISBN-13: 978-1-940631-00-4

Dedicated to One Horse.

INTRODUCTION

Sitting in the last row after rushing to catch a flight, I questioned my decision-making. Everyone and everything was wedged into this small jet waiting for departure in Denver. Glancing across the aisle, I spotted Paul's shaved head as the plane lurched away from the gate.

Even his eyebrow ring seemed annoyed atop the scowl glaring back at me. With his cargo shorts, sandals, and a T-shirt, Paul would stand out anywhere. He wore a large, colorful tattoo on his calf and a few more on his forearm and chest. His broad shoulders were invisible in the penalty box of the middle seat he occupied because I'd booked seats on the plane at the last minute.

Feeling guilty, I resolved not to look at him for the flight to St. Louis. My mind raced ahead, obsessing over the data center rehearsal we had completed a few days earlier. Reality would test our contingency plans in a few hours. Some comfort came in reviewing the *move playbook* in my mind during our flight.

After two hours, the jet landed smoothly and we collected our rental car. Paul, in no mood for conversation or forgiveness, took the wheel while I jumped on the phone with our client. While Paul did the driving to the data center housing the equipment, I monitored the shutdown sequence in a conference bridge call.

After our four-hour drive from the airport to the data center, I wondered if Paul could feel his legs again. His scowl deepened into something intimidating after learning about a problem that would hold up a portion of the shutdown. So upon our arrival, we skipped the normal greeting pleasantries as the staff explained the issue. Temporarily distracted from my seat assignment blunder hours earlier, Paul focused on the issue at hand. His disapproving demeanor never wavered.

Because they'd changed an important administrator password at the last minute, our clients were unable to access a critical system. We huddled briefly while they continued to try password after password.

Unfortunately, they didn't guess the password in time. We couldn't hold the move hostage to this guessing game, or we'd risk a forced power-off of the equipment. A long road trip to get the equipment to its new destination drove the urgency to fix the password issue correctly. As the shot clock counted down, Paul took over and sequenced through the password recovery procedure. He had with him the procedures to recover passwords for all pieces of their equipment.

After solving that problem, Paul's mood noticeably improved. More relaxed, the staff seemed less intimidated by his appearance and grateful for his knowledge. This had all happened before in previous data center relocations. In fact, because it's the first move for most teams, common data center move mistakes often get repeated.

Why I Wrote This Guide

Yes, during my IT career, I've made most of the mistakes I warn you about in this guide *What Everybody Ought to Know Before Moving a Data Center*. I've always marveled that lessons learned are rarely lessons remembered. Many mistakes are driven by urgency; that's understandable. You have a certain idea of what you want, but what's immediate on your mind? An example work breakdown structure (WBS), a budget, a *move playbook*, and/or comprehensive checklists—the *tasks*. Yet it's critical to understand the *process*. It takes more than simply modifying an existing plan. Perhaps a universe exists in which understanding and deep experience is unimportant and "recipes" found online are good enough to complete complicated, sequenced events.

But I don't live in that universe, and that's why I wrote this guide.

Perhaps you have a unique team—one that has a common sense of urgency, makes decisions without running out the shot clock, or shares a common understanding of the process. However, I've never encountered a team that didn't need training, nor a move in which new team members didn't join at inconvenient times, nor an urgent event that didn't require a shift in decision-making.

That's another reason I wrote this guide.

Perhaps you have those rare executives who are experts in every undertaking or who can make decisions without precise data. They are the ones who budget for timely and generous rewards for the deserving. They understand the original budget may be just a guess. They recognize their staff has never moved a data center before and need training and maybe even outside help. But I've met only a handful of these people, so I wrote this guide to encourage behavior that breeds success.

No two data center moves are alike. They can range from complex to straightforward and everything in between. Deciding how to start the planning requires research. But researching online is time-consuming and can produce conflicting information at every click.

For example, there are almost two million results for "data center moving checklist." Even if you could read and synthesize all those results, is that the best use of your time? Professionals today are under immense time and resource pressures. Data center move planning requires learning certain fundamentals and then turning to experienced professionals to avoid costly blunders.

Some project managers build a plan in which everything has to go right and then expend *extraordinary* effort tracking, in minute detail, everything that goes wrong. Others, however, don't take enough time to compose a comprehensive plan. They're in for a rough relocation experience.

Is There a Better Way?

Yes! Moving a data center is a serious undertaking. Any misstep can be costly. Time constraints, budget constraints, organizational pressures, and overworked staff can all contribute to the perfect storm. However, you can expertly navigate through that storm using the foundation laid out in this guide.

Do this: Understand the basic phases of a data center move before you resume your online searching. Form your *move narrative*, establish governance, and make decisions as practical first steps.

I understand the panic you may be experiencing, especially if this is your first data center move. I talk to professionals every day who ride that fear coaster. By providing you with key questions—and answers—the panic will give way to understanding the task in front of you. That's why I wrote this guide. May you put it to good use!

ABOUT THIS GUIDE

Evaluating feasibility early is key; most yelling and screaming about a data center move involves discontinuities between what is *expected* and what is *feasible*. *What Everybody Ought to Know Before Moving a Data Center* is organized to first discuss the concept of move feasibility.

Have you inherited a schedule and non-optional milestones? If so, your move may simply not be feasible, given all the constraints you've been dealt. So tackling this problem upfront will save your strength for the complexities to come.

"Speaking Truth to Power" is one choice you might make to explain the feasibility gap to your boss. (You'll see a section tailored toward common data center move issues and how to get those in power to listen to reason.)

In the section on Data Center Move Phases, the phases are defined and common issues to avoid are discussed. Because cost is the number one concern of clients, the Budgeting phase contains important advice for constructing your cost model.

The *Executive Quick Guide* and *Key Takeaways* are found near the end of this guide to help busy executives orient quickly.

Finally, the *Self-Assessment* section helps you decide where you might need outside help.

Where I live, springtime in the Colorado Rockies sends car-seeking rocks hurtling down valley roads. While traveling the winding passes, you often can't see these hazards in time to avoid them. Like the "Watch for Falling Rocks" signs, this guide warns you of the hazards awaiting your data center move journey so you can avoid those rocks.

Isn't that your goal? To move your data center without getting hurt?

The Most Important Point

Participants enter a data center move with different levels of understanding and a different sense of urgency at different times. Almost no one starts at the beginning and sequences through a linear checklist to achieve a happy ending. So no matter what stage you find yourself or the degree of panic you feel, take time to understand all of the data center move phases.

Why do I point this out? Because the most important part of a data center move is *you*. You're the one who curates the available information and adapts it to the circumstances of your move. By grounding yourself in the fundamentals, you can effectively manage the urgency, people, and process required to complete a data center move. And avoid the rocks!

What's Not in This Guide

It's equally important to explain what is *not* in this guide. You won't find full-blown work breakdown schedules or past actual data center move budgets. In fact, no easy buttons exist within this guide. If you prefer to copy a spreadsheet or a Microsoft Project plan, then you won't find that here. But to learn the principles and practical considerations of your imminent data center move, start reading now.

Application inventory and dependency mapping is another complex area that's impractical to address here. This guide explains why it's complex and potentially expensive, but there is no easy button here, either. A successful move comes from understanding the complexity, constructing a meaningful plan, augmenting your resources, and onboarding others at inconvenient times.

Simply stated, success springs from you first. It requires you to invest preparation time for all the complexity you will encounter.

Let's start by exploring the feasibility of your firm's data center move.

TABLE OF CONTENTS

HOW WILL YOU PUT HUMPTY DUMPTY BACK TOGETHER AGAIN?

WHAT'S THE ANSWER TO THIS QUESTION?

You dismantle your entire data center, move it to another location, and re-assemble the pieces. And like the nursery rhyme, you fear that "All the King's horses and all the King's men couldn't put Humpty together again."

Exactly what is a data center move? Think about your move in two ways:

- Physical—The computers and supporting equipment are physically moved from the origin to the destination.

- Logical (sometimes called Migration)—Rather than moving the physical components, the data and applications are moved from their original place to a new location. Virtualization is a technology that enables physical servers to migrate as files and will be useful in both physical and logical moves.

However, nothing about your move forces you to choose any one method over the other. Selecting the best method for the applications running your business often means a hybrid move involving both physical and logical techniques. It's wise to know your applications well before making a judgment on how to move them. Because many choices exist for the destination, your applications won't necessarily end up at the same destination. (These choices are covered in Site Selection.)

No typical move defines all the choices, but as you explore this guide, you'll be able to discern which elements are important to your move. You'll also learn a way to document your choices. But before you worry about reassembling Humpty Dumpty, determine if the move is feasible.

The First Question to Ask

The first question you might ask isn't about cost (addressed later in this guide) but "Is my data center move feasible?" Can you predict your chances of a successful move with the information you currently have? Therefore, you would do a feasibility analysis—*the objective study of strengths and weaknesses so one can render an opinion on the success or failure of an endeavor.*

While that definition is a mouthful, it's also important to understand what a feasibility analysis is *not*. It's not an exhaustive, time-consuming exercise that could end in a tie. After all, the analysis is not your primary goal; it's moving the data center. So you don't want to expend extraordinary effort and resources to arrive at a precise answer. But you do want to examine these factors:

- Applications
- Components and elements
- Turnovers and pitfalls

Review Applications

Applications run your business, so their availability is important. This makes application downtime an important metric. Certain applications depend on a range of important services provided by other elements of your infrastructure.

Applications and their dependent services all require the hardware be configured correctly. In turn, hardware requires network connectivity, power, and cooling to operate. A properly equipped space such as a data center houses all of this. As well, it includes monitoring systems to alert for any anomalies.

When you endeavor to move an entire data center, it can feel like a theatrical production that relocates from one "city" to another. You must consider all of the behind-the-scenes components before you can settle in.

This level of complexity requires people: a combination of service providers, in-house staff, and equipment manufacturers. Keep in mind that *people* are the most important element of your data center move—starting with you!

Visualize Components and Elements

To think about your move and its feasibility, take time to visualize each component and the elements contained within each container. The data center is the physical room that contains power and cooling systems. Computer servers in that data center (hardware) host applications that run your business. All of this complexity takes supporting staff to design, configure, and operate. Behind every application on a desktop or mobile phone is a data center that contains the services and content.

Not surprisingly, many project managers accept untested assumptions without verifying them. What pitfalls await? Untested restoration of data backups, unlabeled inventory, and systems long since out of support could be true for your organization. Have you seen this movie before?

If victory loves preparation, then an unmitigated disaster is thrilled when there's haste in building a plan or presentation and when the budget for moving a data center is a guess. You may have a pressing urgency forcing you to search for something very specific. But you can appreciate there's no magic way to accommodate some seekers without reading their minds. Because I don't have that skill, contact me instead with specific questions.

Most initial client consultations take a few hours over the phone. The anxiety IT managers express in those first few minutes is understandable. Why? Because many firms try complex projects themselves and then reach out for help only when they've run out the shot clock—an analogy borrowed from basketball. By then, all the choices left are expensive and the panic remaining is quite palpable. Let me explain the concept of the shot clock.

In the game of professional basketball, a turnover results when a team fails to take a shot within the 24 seconds allowed. A similar kind of built-in urgency is often missing from today's leaders. They fail to see the consequences of their unwieldy and elongated decision-making.

Leaders including coaches have to articulate the organization's strategy and make critical decisions in a timely manner. In basketball terms, that means leaving time on the shot clock for their team to execute. Do you expect to make any progress if you consistently run your decision-making shot clock down to zero?

Due to your own inaction, what turnovers might you be committing? Consider these:

- You over-rely on consensus instead of identifying the true decision-makers or exerting your own authority.
- You fail to understand when the possession arrow points to you and others are stalled awaiting something you promised.
- You are blocked by the indecision of people above you, perhaps because you fail to articulate your needs and their urgency.
- You passively let crises dictate priorities instead of systematically preventing emergencies.

My advice? Leave time on the clock for your team to execute well.

For a data center move, the right time to engage more resources might be well before the shot clock has reached zero. Would you fly an airplane for the first time and then, only after it crashed on takeoff, seek an experienced pilot? Professionals are frequently viewed as expensive luxuries, but that doesn't make going with amateurs cost-effective by any stretch.

Avoid Turnovers and Pitfalls

How do you avoid pitfalls when you are moving a data center for the first time? For example, imagine you're driving blindfolded with only one instrument available. A special talking compass blares warnings along with the direction you're heading. “Uncertain you are going north” or “mostly certain you are going north,” it warns. Those two different facts influence how hard you press the accelerator.

It’s the same with a data center move. If you are mostly certain about your plan, you can go faster without guessing or restarting the analysis. If the uncertainty is high, then pressing hard on the accelerator will earn both a spectacular crash and an expensive repair to restart your journey. Apply this concept of an uncertainty gauge—a method to quickly determine which areas are guesses and which are fully understood.

Assigning a number scale informs your priorities and keeps you focused. Sadly, we expend too much effort on what we know while ignoring the tasks necessary to identify the unknown hazards. How often have you experienced endless meetings about topics that have been flogged to death only to learn critical items remain untouched?

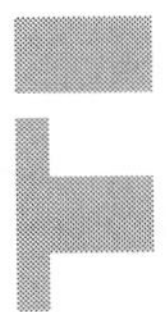

APPLY THE CONCEPT OF AN UNCERTAINTY GAUGE

How do you rank the information you have against an uncertainty scale?

A practical method using a spreadsheet is to assign a number scale to uncertainty (1=certain, 2=somewhat certain, 3=uncertain). Then calibrate your uncertainty gauge against your collected facts. (Look for an example later in this guide.)

Since data center moves are not normal events for most staff, it's certain entire segments will be untouched. Unlit by discovery or discussion, these neglected items crawl out of the shadows to disrupt most moves.

The discipline of feasibility analysis requires you spend time outside the streetlight of your own understanding. You turn to others who can quickly illuminate the dark areas of your plan.

Start Your Feasibility Analysis

Still, having a basic understanding of a data center move for yourself is equally important. These eight questions will help you start your feasibility analysis:

1. Has the destination been chosen?
2. Do you know your budget? What is it?
3. Have you prepared a written move plan?
4. Have you prepared a risk and contingency plan?
5. Have you assigned responsibility for each phase of the move?
6. Have you determined the impact of downtime to your business?
7. Do you know who needs notification internally and externally?
8. Do you have a move timeline established with lead times documented?

I ask these eight questions in every initial consultation with my clients. The answers quickly tell me if the organization is truly ready to move its data center. If you've answered "yes" to all of these questions, that means you've completed a substantial amount of work to arrive at this point. It also means your firm is serious about dedicating resources to such an important event.

Understanding the fundamentals in this guide is an investment you'll leverage as you plan your data center move and execute your plan.

CAUTION: If too many items are undecided and you have a sinking feeling far too much effort will be required, then put aside this guide for now. Get ready first. Your data center move involves hard work with no online easy buttons.

Mastering Uncertainty in Your Data Center Move

Along the journey of your data center move, uncertainty will be your constant companion. Use a process that is helpful and repeatable. Not every uncertainty rabbit needs chasing with the same priority. Nor do they all need chasing by *you*. However, it's imperative you keep score. Is your uncertainty growing or decreasing?

The best way to handle uncertainty is to document what you know now. As you learn more, your *move narrative* will take shape. A *move narrative* provides a place to collect the known and the unknown—a place for your fears and hopes to get calibrated by your uncertainty gauge.

***MOVE NARRATIVE* QUIZ**

To get started, take this "move narrative" quiz:

1. *What components are moving?*
2. *When are they moving?*
3. *How are they moving?*
4. *Who is doing the move planning, execution, and cleanup?*
5. *Do you have a budget?*
6. *Do you have a written move plan?*
7. *How much downtime can you tolerate?*
8. *Do you have a physical inventory?*
9. *Do you have an application inventory?*
10. *Do you have current, documented network drawings?*
11. *Do you have equipment elevation drawings for the destination?*
12. *What specialized equipment do you have requiring extra attention?*
13. *Do you have a plan for decommissioning the origin?*
14. *Do you have a plan to celebrate your success?*

Don't worry about incomplete answers. By doing your best to answer these questions, you've taken the first step to documenting your move narrative.

Why is a *move narrative* important? Because it helps you organize your initial thoughts. It also:

- Becomes the historical record along your journey, allowing others to quickly understand decisions no matter when they come on board.

- Shapes your actions and narrows your focus, highlighting the elements of uncertainty that need attention.

- Informs your initial communication to convey to your vendors and staff members.

- Minimizes chasing out-of-sequence activities that might be re-done anyway as you learn more.

As you learn about the data center move phases, revisit your *move narrative* often. By the end of this guide, that data center *move narrative* will warn you about hazards in a far better way than a collection of online tips and checklists.

The Value of a Move Narrative

There's no magic about writing a *move narrative*. You simply write the story of your move before it happens. Who are the characters? What skills do they possess? What road defines their journey? What happens along the way? What limits their choices?

In the beginning, company leaders change their minds as data informs their previously held assumptions. Therefore, a *move narrative* can change frequently before consensus is reached. That's fine. Do you have time to boil the ocean of the possible move plans? Probably not, but you must also convert at least one *move narrative* into a completed plan. Integrate the changes before the massive effort begins. However, don't wait too long to expand one of those narratives into a draft plan for enhanced scrutiny.

Move Narrative Example #1

Consider what seemed like a simple *move narrative* at the time for my St. Louis client:

> *"Move the accounting systems and file servers from the origin to the destination by truck. Internal staff will shut down, tear down, transit, and reassemble the equipment. The tolerated downtime is twenty-four hours. The origin will be decommissioned following the move. There is no business continuity location. Out-of-country manufacturing depends on the file servers being moved. Those manufacturing locations are time zones away from the Central time zone, one being in Asia and one in Mexico."*

Was this move feasible, given these constraints? Not even close. The staff had not properly accounted for these time-consuming elements:

- A full backup of all systems constrained by one backup device and associated software.

- The time to shut down and tear down the equipment.
- The packaging time to secure systems properly for transit.
- The proper calculation of the route to include transiting through St. Louis during rush hour.
- The slower drive time required for a truck compared with a car.
- The team drivers needed and the refueling stops required.
- The reassembly time taking longer because no pre-staging of rack furniture would be done due to cost.

If they had missed these elements, what else could be missing? Clearly, the *move narrative* had suffered from the warning stated earlier: Too much effort expended on what they knew while critical items were left undetected. Undersized bandwidth at the destination meant immediate performance issues following the move.

The time zone differences of two additional locations indicated more understanding is needed. Plus how would this move affect manufacturing? Many fears proved accurate.

I suggest writing out the *move narrative* in a document instead of a spreadsheet. That way, its value goes beyond generating cost elements for a budget. When confronted with their *move narrative*, most executives immediately solve for what's missing, which is a desirable outcome. Extracting this information early means avoiding expensive blunders later.

As the *move narrative* unfolds, it's possible to challenge the stated constraints. In example #1, how was 24 hours chosen as the allowed downtime? Probably an arbitrary choice or a wild guess so they'd have a number for calculating moving expenses. But if constraints go unchallenged, then how will you contain costs? And if the *move narrative* does not even document the constraints, then how can "ghost" limits be challenged at all?

Move Narrative Example #2

I received this *move narrative* from a client who had already set the move date but hadn't chosen the destination:

> *"The move will involve 23 racks of equipment. Some are moving physically and some will require only the data be moved. We haven't chosen the destination yet. We have a deadline to exit the origin in six months. We don't know the downtime allowed but believe it will vary by system moved. We believe that moving over a holiday will be the least disruptive to the business. We need to split the move into several move events because we feel it will be risky to move everything at once. How much will this cost?"*

Working backwards, let's make a few observations.

Splitting the move. This is a good strategy, but it's also highly dependent on several constraints. To answer a move split, you need to know how applications are grouped and what resources they consume. For example, if you move a storage array *all* the applications use, then you either must move *all* the applications at once or find a second storage array to host applications as you move them.

This example ignores the fact that unintended performance results occur when applications are moved without regard for how these applications access their resources differently after their move.

CAUTION: A common pitfall is performance degradation when applications at the destination reach across the slower Wide Area Network to access resources residing at the origin.

Moving over a holiday. Treat this as a red flag to be avoided. (More will be discussed later.)

Not understanding the business requirements of downtime. This is more common than you might think! It's imperative you hunt this rabbit and hold business units accountable for agreeing to downtime allowances. If they want zero downtime, you know your next stop is to find out who will pay for that constraint. Know this: Zero downtime is not free. So weigh the cost of the complexity of zero downtime versus using certain windows of acceptable downtime. This real-world tradeoff is worthy of more than one thought experiment to solve.

Not deciding destination. Every data center move is different and many destination choices are financial decisions based on economics. Take extra care to avoid using up the entire shot clock on a destination decision. You have a long journey after determining the destination, so increase the urgency and your attention to this step.

Combining the physical and the logical move. This is a common choice. It also indicates you need to understand all the complexities for accomplishing this objective. Data transfer over wide area networks never operates at the theoretical maximum. Know there's always a production penalty for simultaneous data transfer while still operating your business on the same network.

Stating the number of racks. The first statement notes about 23 racks. This indicates more detail is required. Better to reduce the 23 racks of equipment into a physical inventory of make and model as soon as possible. Inspect that information and determine which machines might be off of maintenance, which can be retired, which might be duplicated, and which can be virtualized.

After reviewing these examples, you can see how a *move narrative* embodies this structure:

- What is moving? (Quantities of what type of applications and systems?)
- What decisions have been made? (Destination, transportation, and resources used?)
- What tolerated downtime exists? (How is zero downtime accomplished?)
- What are the other constraints? (Milestones, budget, business drivers?)

Focus on these basic questions before launching a detailed project plan. Make early basic decisions knowing they can be revisited to alter the course if necessary. Remember, unmade decisions are like shots not taken before the clock runs out.

CAUTION: Expect that 100 percent of the unmade decisions will be wrong and wasteful. The *move narrative* helps develop a skill you'll appreciate later in your move. And accept that nothing works perfectly

---

Long before Paul and I boarded that flight to St. Louis, we had created and refined the client's *move narrative*. In fact, we cussed and discussed the *move narrative* until a measure of consensus solidified our choices. It quickly became a *move playbook*—one that gets rehearsed as discussed later in this guide.

HOW TO SPEAK TRUTH TO POWER

CAREFULLY OR NOT AT ALL

Something that gives pause to most professionals may manifest early and never be completely resolved throughout your move—that is, speaking the truth to people in power. Someone's fear prevents telling the good, bad, and ugly truth to those in charge.

In managing your data center move, schedule slips and cost overruns are the most feared news you might need to deliver. Although this fear depends on the corporate culture, no one likes to deliver bad news. Do you want the top executives to blame you for something beyond your control?

You might already be experiencing fear with your data center move—for example, fear of being blamed for impossible deadlines or fear of losing your job for not executing a flawless move. That's normal. On every data center move engagement I've done, fear of job loss has been expressed both from staff and management alike.

In St. Louis, Paul and I encountered one particularly acute case of fear paralyzing an entire staff. It involved a complex data center move and missed milestones. Inconsequential political infighting was wasting the shot clock; important technical work had gone unmanaged and unwatched. After some digging, we learned that, while much of the staff knew of the impending disaster, the management team did not. Why does this happen? The answer is as much a statement of human behavior as it seems incredible.

The technical staff wanted the management team to miss the non-optional deadlines. Why? Staff members had correctly calculated that a missed data center move would result in a few management dismissals. Willing to sabotage the important data center move by remaining silent, they planned to accomplish something that years of employee opinion surveys had not.

It should not surprise you to hear that data center moves expose existing organizational dysfunctions as well as perplexing individual behaviors. These behaviors don't get solved through clever phrases on social media messages, nor do they disappear as everyone pulls together for a common goal. Who finds themselves during a data center move sitting around the campfire holding hands and singing songs?

If you hurriedly pen an e-mail that goes to the top of your corporate food chain claiming the sky is falling and everyone except you is to blame, are you speaking truth to power? Or is this simply a defense to escape *why* you should not be held responsible?

Speaking truth to power requires a lot more work and significantly more courage than hasty, cover-yourself e-mails. To examine this, let's deconstruct three common reasons:

- Your truth may not be true at all.
- You may have misidentified the power.
- You are the wrong messenger.

Your Truth May Not Be True at All

In the complex undertaking of a data center move, technical staff may be blind to the economic drivers and business imperatives. That's not to say they are *always* blind. Their confirmation bias may lie in believing that technical solutions to problems are the singular answer.

Confirmation bias is when you seek or recall only the evidence that supports your beliefs. It goes almost unchecked with information technology staff members; they naturally gravitate to their strengths and discount or overlook elements that are unfamiliar. Because data center moves are not normal events for most staff members, they may be unfamiliar and fall victim to excluding important facts through confirmation bias.

Consider what happened when the technical staff recommended buying a new but highly expensive storage array to substitute for moving the old, legacy array. They spent extraordinary effort to justify the purchase and were flatly refused by the executives. Using the shot-clock scarcity to attempt to justify this choice, they realized no time remained to prepare the legacy array for a complicated move. Because of confirmation bias, only this singular solution was pursued.

It certainly appeared that a failure to buy the new array jeopardized the schedule. And even though the staff failed to justify it, staff members reasoned that management would be blamed for making the wrong decision resulting in the schedule slip—a perverse interpretation of a false victory.

Paul knew a different truth. Pursuing a parallel strategy with the vendor of the storage, he had already secured professional services to move the legacy storage array on schedule in case the new array could not be justified. What's the lesson here? Before you speak truth to power, make sure you source more than one alternative for making progress on your complex move. Strive to find alternatives then talk to experts who can argue for or against them. When you speak truth to power, speak *all* the truths and accept the decision. Most important, *actively support* the decision with your actions. If you sabotage a data center move because you disagree with a decision you lost, it can undermine morale during a difficult data center move. And it's unprofessional.

CAUTION: Honestly answer this question: Is your truth colored too darkly by your confirmation bias?

Misidentifying the Power

The ubiquitous organizational chart identifies the power structure in most companies. It's a roadmap for political escalation and infighting. And often, it's the correct escalation path for conflicting decisions—but not always.

As discussed later, a data center move typically has a governance body to facilitate critical decisions. Moving these decisions up the hierarchal organizational chart can be frustrating for a host of legitimate reasons. For example, while being expected to perfectly understand all the complexities of a data center move, executives still have to focus on running their firms. It takes them time to understand the entire technical nuance of a critical decision.

Because *you* are living and breathing the data center move, it doesn't mean it's top of mind or high priority for the executive you want to influence. One chief information officer (CIO) presiding over a lengthy data center move described his dilemma this way: The data center move rehearsal exposed serious concerns about the state of readiness for affected departments. Vice presidents of these business groups needed to receive an unpopular directive about having to rehearse manual procedures before the data center move.

In today's data environment, no one wants to resurrect the paper-and-manual procedures, so the resistance to practicing these processes is great. Predictably, vice presidents wanted a flawless move so they can avoid the rehearsal work. Does the CIO escalate this reluctance and risk antagonizing a power group necessary for corporate harmony? Does he instead pray for a perfect data center move and take on all the risk of a blunder?

Be wary of an intractable situation—until you realize your *customers* are the true power. In this example, the data center move might affect them if manual contingencies aren't practiced. Armed with that insight, I challenged the CIO to contrast the inconvenience of doing a rehearsal of manual procedures with the cost of repairing negative outcomes for the firm's customers. Realizing that technical staff members weren't being lazy—rather, they wanted to ensure continuity for customers should the worst case occur—the VPs changed their minds.

The lesson? Identify the right recipient of your message (and it might not always be the next person on your organizational chart). Keep in mind that those ultimately affected when a data center move goes wrong are your customers.

Always find ways to represent them in your decisions before automatically escalating up the organization chart.

Being the Wrong Messenger

A data center move does not suddenly make you the subject matter expert on diplomacy and decisions. In practice, it's hard to give up being the messenger of this change. Why? If you're being honest, you *truly want* the credit for it!

Data center moves expose technical staff to executives, often for the first time in their careers. Who doesn't want to maximize that kind of exposure? This powerful lure can get in the way when something important needs to change. Suppose you're not the right person to do the changing, but you attempt it anyway.

Here's what I've observed. Practically no one gives up the opportunity to be the messenger when boosting one's reputation is imminent. Given this situation in your organization, I suspect you'll be no different. When you lament your failure to speak truth to power, consider that although your truth is correct, you are the *wrong* messenger of that truth.

What's the lesson? If you truly care about the outcome, then don't give up. Adjust your tactic and find another messenger—the right one for your purpose. With time being in short supply during a data center move, expect confrontation to be inevitable and necessary. Therefore, any time spent positioning yourself as a messenger for your own credit is actually time you're wasting.

When Paul speaks truth to power, he often removes his eyebrow ring to avoid any condescending judgments about his style. This is a reality he understands. Perceptions color the ability of others to receive your truth. As Harry S. Truman is quoted as saying, "It is amazing what you can accomplish if you do not care who gets the credit."

CAUTION: Be aware of your motives. Do you want the accomplishment or the credit? If you determine you want the accomplishment, seek another messenger and get on with the hard work of a data center move.

Treat Power with Respect

Too often, a condescending tone can permeate attempts to persuade executives to choose one option over others. Technical staff presume the only worthy argument is a technical argument; they quickly become frustrated when executives do not grasp its significance. However, arguments based completely on technical data do not automatically make them the correct choice.

Knowing Paul and I were competing for a large multi-year data center move project, it was time to defend my proposal in person. Paul agreed to remove his eyebrow ring and forego his shorts and sandals for a black suit, white shirt, and polished black shoes. As we waited in the lobby, a competing team completed its presentation and then exited past us. A perfect parade of a dozen men and women precisely groomed talked excitedly about the session they just finished. I was worried. Were we outnumbered and outmaneuvered by this confident team bubbling about the impression they just made?

We then stepped into the conference room to see five people sitting on one side of a long, narrow table. We were the last of three presentations that day, and we could see the panel members were clearly tired. Worse, they didn't even look up from their laptops.

The question and answer format they used seemed rushed. Were they even listening to our responses? Then came this final question: "Is there anything you want us to know about your proposal?"

I've heard that question many times in my consulting career. My answer can make or break a selection decision. So I decided to speak truth to power. I could tell my audience made up of the CIO and his four managers had varying technical depths. I took a deep breath and began speaking slowly.

"Your data center move is not feasible, and here's why." Then I explained all the non-technical reasons why it could not be completed according to the milestones they wanted.

"Do not hire anyone, including us," I stated. Then I explained why they needed to make a few basic decisions first. I also said why their methodology seemed inappropriate for a complicated data center move. When we walked out of the room, I remember how somber I felt and the worried looks on their faces. What a contrast to the jubilant parade of the previous presentation team.

Lo and behold, against two impressive and formidable competitors, we won that bid. After the engagement completed almost two years later, I sat down with the CIO to learn why. In a word, "Respect."

He said that all the experts who presented intimidated his management team by telling them they couldn't possibly understand how technically complicated the data center move would be. They felt frustrated by the technical staff speaking in technical riddles. As a result, it stalled effective internal planning. In contrast, my firm treated them with respect while still telling the unpopular truth about their data center move. And we did it in understandable terms.

The lesson? Speak truth to power with respect for your audience. Help people understand the issues in their own language. Listen carefully to understand their circumstances.

Remember the example of the staff sabotaging the data center move in an effort to remove management? In almost every case, one or two individuals try to sabotage their data center move. But an entire staff? If you get wind of that, urgently address it. That's big. Realize that a staff with no confidence in their management team is a major dysfunction. A data center move guide simply can't resolve that issue satisfactorily.

After being confronted with this show-stopping problem—that is, fix the organization *before* the data center must be moved—I offered the CEO this observation: "Your organization is broken. I can't fix it in the time I have. This is what it will cost to move the data center by replacing the entire staff and the management team." Then I presented an extremely large number with charts that supported it.

I also explained the additional disruption costs of losing subject matter experts and the collateral damage to customers. They were beyond shocked at the magnitude of the problem.

Actually seeing costs assigned to the dysfunctional plan motivated the CEO to dig deep, enlist other business units, and find an alternative to wholesale replacement.

I won't tell you the problems in this situation were solved painlessly. But the direct and continued participation of that CEO did result in a happy conclusion. Most important, a dysfunction that had gone purposely ignored was finally addressed in an honest, meaningful way.

CAUTION: Data center moves will expose both the best and the worst of your firm, so be prepared for both realities.

The Problem of Misalignments

The skills to accomplish your mission of moving a data center simply may not exist in your organization. I call that a *misalignment* of skills and mission. And while this misalignment may be crystal clear to you, it's not necessarily clear to the executives you work with.

Trying to convince your executives this misalignment exists often results in them perceiving you as a negative person—someone who views the organization as unskilled and inadequate. And in attempting to make your point, you might only dig yourself deeper into this misperception hole. That could put your audience on the defensive. You risk them closing down to your suggestions.

Remember, misalignments exist in data center moves for many legitimate reasons. Your job is to communicate those misalignments to executives effectively. But how?

First, recognize that executives wrestle with misalignments every day throughout their organization. They evaluate support structures, skills, and processes against the changing needs of their firms. You could tell them they have a fundamental misalignment. If you expect an immediate fix so the data center can be moved, you'll wait for the sound of crickets. Nothing is happening!

Avoid boiling the ocean of company culture and processes by staying on topic about getting the data center moved successfully. Don't make it the critical item to solve before you even get to the data center move. Yes, executives have seen that movie before. They know it's not realistic to transform culture and processes solely to get their data center moved. They understand the law of unintended consequences. They know that distractions burn money at an alarming rate while not fixing the problem at hand.

If skills are deemed the problem, then give the executives options and plans for mastering the needed skills. Which specific skills? For what duration? Be certain you've completed your work and presented *all* issues and recommendations in an actionable way.

CAUTION: One of the hardest things *ever* is to change a company's culture while it's in the critical path of solving a specific problem. Don't make that mistake. Instead, define exactly what problem you need to solve. What are the alternatives you've considered? Which ones haven't been looked at yet?

SUMMARY GUIDE AT END

*Look for an **Executive Quick Guide** at the end of this guide. It's for executives who don't have time to read everything in this book. Regard it as critical insight in compact form for the complexities awaiting your data center move.*

After you soak in these concepts of feasibility, the *move narrative*, and speaking truth to power, let's move on to the phases governing most data center moves.

DATA CENTER MOVE PHASES

WHAT STEPS ARE INVOLVED TO SET UP YOUR MOVE?

Many of my clients want to avoid learning about the data center move phases. They prefer to skip directly to knowing how much the move will cost and when it can be completed.

You may be similarly tempted to skip the step of fundamentally understanding move phases as you rush to arrive at the answers to those two questions. However, understanding move phases helps you manipulate them to control costs or accelerate progress. Consider that a small investment compared to the cost of making avoidable blunders due to ignorance of the process.

Realize there is never *one* answer about how much time and money a data center move takes. Every answer depends on how you execute your move plan. If you make avoidable mistakes, your costs will soar. If you fail to optimize certain steps, your timeline will elongate. As you build your *move narrative and* your understanding grows, you will find additional opportunities to contain your costs and improve your schedule.

Breaking down your move into several phases helps you begin the planning process. The phases of your move consist of:

- Governance
- Budgeting
- Site Selection
- Pre-Move Planning

- Teardown
- Transit
- Arrival
- Re-Assembly
- Post-Move

Precision in Communication

Pay attention to the lexicon or vocabulary of a move as you read this guide, knowing that a common language enforces precision in communication.

In most firms, precision in communication is a daily struggle. Every group uses terms and acronyms foreign to outsiders—and that's just one aspect of lacking precision. Be sure to use a common language with defined terms. These suggestions might help:

- Designate an authoritative source (e.g., your Communications Department) for all of your move-related communication.
- Use a formal process for collecting move-related questions from everyone.
- Answer common questions and distribute Frequently Asked Questions to newcomers.

- Over-communicate, not under-communicate.
- Be direct! Leave out the wordiness.
- Be coherent! Don't make readers connect the dots; do it for them.

Building a shared vocabulary early helps everyone who comes on board later. For example, "origin" and "destination" are terms designating the starting and ending locations for your move.

Document the terminology your organization will use with internal and external resources to flatten the learning curve for all—no matter when they show up to help. If you can anticipate the effort that the onboarding process can steal from a data center move, it helps build a contingency against that threat. *This guide is a tool you can use for shortening the time it takes to bring new people on board.*

Your data center move may have been funded before anyone knew the actual cost or set the budget. But Budgeting doesn't come next. Let's discuss the concept of governance—how decisions get made and by whom—before we cover Budgeting (despite the obvious chicken-and-egg paradox).

GOVERNANCE

WHO ARE THE DECISION MAKERS?

Most large firms have some form of governance in place. However, these structures may be inadequate for data center relocations. Still, getting governance right is critical for a well-planned and well-executed data center move.

Governance is *about who makes the decisions and how those decisions get made*. It focuses on explicitly declaring the decision rights and the accountability for the project. It then adheres to a disciplined monitoring process. If you involve every single person in the data center move decision process, you'll find the price for excessive consensus is the opposite of agility, efficiency, and accountability.

I remember a complicated move in which avoiding *individual accountability* was the sole objective for the governance team members. What is the first hallmark of this governance style? Excessive consensus. This can be concerning because long decision cycles and baffling outcomes occur when everyone avoids personal responsibility for a data center move. Although entire books have been written about governance, few address the problem of shot-clock violations to avoid responsibility.

CAUTION: The best way to waste time is to assemble responsibility-avoiding participants who can't make independent decisions in a timely manner.

Starting the Governance Process

Get started right with these four simple steps:

1. Identify the stakeholders and their decision-making authority.

2. Add an external advisor to provide a catalyst and unbiased viewpoint.
3. Publish the decision framework for discussion.
4. Dedicate the resources to allow regular governance meetings.

As you make your plan, be aware of these common mistakes:

- ***No agenda***—Ad hoc, freewheeling meetings are huge time-wasters. Establish a precise agenda for every governance meeting.

- ***No urgency***—Governance meetings are not an excuse to slow down a project. With real-time collaboration tools, there is no reason for this type of organizational coefficient of drag.

- ***Politics***—A data center move is not the time to grind political axes. It takes a strong leader to minimize any politics. Instead, focus your resources on the complexities of the task rather than its politics.

- ***No responsibility***—If governance doesn't want the responsibility for the data center move, consider appointing a data center Move Commander instead.

A problematic data center move benefits from introducing or revamping governance. Adding an unbiased, external adviser can give you the kind of reality check needed to move the project forward. Especially if governance is a new concept in your firm, seek expert help. An external advisor could be someone in your own company not directly involved in the data center move. That person's fresh perspective within a legacy governance structure can counter a range of avoidable decision-making problems.

Imagine you are building a house with the only tool you know how to use—a rock and nothing else. Now, suppose your neighbor builds a house with a full selection of modern tools. Can you predict which house is built better?

Consensus decision-making is that old rock of available techniques—the overused tool—despite the availability of more powerful and effective strategies. Like the homebuilder who has only a rock, a leader making decisions only by consensus will fall quickly behind other firms who understand and practice the alternatives.

It's easy to fall into the consensus trap. People tend to be taught early and often to involve all members of the group in decision-making. Presumably, better decisions arise from gathering everyone's input. Yet consensus decision-making only makes the *acceptance* of the decision easier because everyone involved had a voice. It doesn't necessarily make it the *best* decision or the *right* decision. And it certainly can ensure it's the *slowest* decision.

Still, leaders practice consensus decision making every day as if it's the only tool around. But it isn't the sole trap awaiting the important process of decision-making. Consider these:

- First impact trap
- Status quo trap
- Confirmation bias trap

The first impact trap

Your immediate reaction to the first piece of information or the first idea received can carry a disproportionate weight. It blocks evaluating all of the information available with the same enthusiasm you had for the first idea. Not surprisingly, the first impact trap colors all subsequent information.

The status quo trap

You may have built-in bias deciding to do nothing even when better alternatives exist. Risk avoidance, work avoidance, and cost avoidance are common illusions reinforcing the status quo. Remember, there's always a cost for doing nothing. For a data center move, doing nothing can leave you with only expensive choices to salvage a schedule because you failed to act out of fear or responsibility avoidance.

The confirmation bias trap

When people suffer from confirmation bias, they seek information to support their beliefs, and they discount or even ignore information undermining those beliefs. Staff diversity helps overcome this confirmation bias.

Do not assume your governance participants understand effective decision-making techniques. You don't want them to fall into these traps or doom the data center move to using only the old rock of consensus decision-making.

GOVERNANCE AND FAILURE

Why is governance that's deemed so important often so neglected? For insight on this question, I turned to Charles Nelles. He's an enterprise ninja in IT Operations and Service Management who has led high-profile projects for two decades.

"Charles, why does governance seem to be a universal failure?"

*"You name it: executive sponsorship, corporate culture, underestimating the effort, and failing to stick with it. In my opinion, governance fails because of people. It's about doing the **right** thing, and most firms don't prioritize that over doing the **expedient** thing."*

"Can you give an example?"

"Picture a conference room table with people sitting around it. They're all looking down at printed slides or their mobile devices. They're wringing their hands and

racking their brains in a desperate effort to avoid being the one to make a decision.

"They ask questions but not the kinds to eliminate possibilities or narrow decision paths. Those questions show their concern or their knowledge instead of driving toward an outcome. Then they tell me what they are worried about such as 'if I don't have at least six hours to complete this heroic action, I can't guarantee we will have the system up and ready.' In this way, they're communicating that 'while I am a genius, I am not to be held accountable if things go wrong.'

"I call this the puppy circle."

"The puppy circle? Help us understand that analogy."

"Imagine a puppy with a bucket stuck firmly on his head. Now imagine a circle of people around the puppy discussing how horrible the situation is and how worried they are for the puppy. But no one is reaching for the puppy or the bucket. In many governance meetings, no one reaches for the puppy. The first responders are absent in a puppy circle."

Charles also stated that effective governance should follow a design. He recommends three design elements be included from the start of the process:

1. ***Speed to Decision****—Whatever the structure, assemble the decision makers (or appropriate delegates) quickly and make the call. Wasted hours or days directly affect your cost and chances for success.*
2. ***Proper Subject Matter Expertise****—Have enough knowledge in each key domain to make smart decisions with limited information. This combination is not always easy when the technical world meets the business world—as in a data center move.*
3. ***Respect and Buy In****—Regardless of who or how many the true decision makers are, engage those in the larger group at some level. Continue to manage their perceptions of the decision-making. Everyone who can add to the success of the move must feel like a valued member of the team who has a level of influence over the decisions. If you lose the 'huddled masses' who will actually do the work, you are sunk.*

In the second design element, governance may lack deep technical experience to understand the collateral damage of decisions. At the same time, the technical staff lacks the business experience to commu-

nicate understanding. The two camps are rarely in the room together.

To resolve this, Charles said, "This is the paradox of Information Technology in general. Let's face it; a tight coupling of business and technology minds is the nirvana most firms seek. But it only takes one person to cross the line into the other's territory—one person with influence over the project to understand this paradox. That can sometimes be enough. Pull the puppy out of the bucket first and then use action as an example of the behavior to reward.

"Most of all," he cautioned. "Don't engage in useless activity. Those aren't my words; they come from Miyamoto Musashi in ***The Book of Five Rings****. In my mind, that is the single purpose of governance: to ensure we do not waste precious company resources and put the success of the company at risk with useless activities." Words to the wise.*

Let's look next at a phase that's essential to master—Budgeting.

BUDGETING

HOW MUCH WILL THIS MOVE COST?

Budgeting for a data center move can be frustrating because, at the beginning when information is scarce, precision is elusive. Plus, no single budget number exists. Rather, the budget is a function of your *move narrative* and the path you are expecting to follow. That path will change along the way—along with your numbers.

However, pressure from management to "get me some numbers" often results in missing important budget details. Putting your faith in "standard" per-square-foot budgeting numbers can leave you without a realistic view of the real costs in money, time, and internal resources for your firm. What's more, setting an unrealistic budget expectation means your lead time, coordination, and notification may be inadequate. Worse, the project could be underfunded due to willful budget under-sizing in an effort to gain approval. This can lead to costly shortcuts and ill-advised decisions.

Assumption Errors Cost Money

What is the most common assumption error made during budgeting? *Assuming your cost model is complete.*

Even small assumption errors can be quite costly. Assuming, for example, moving on a holiday will cause the least disruption may not only increase your costs dramatically; it could also interject unnecessary risk into your move. Critical support resources simply may not be available during holidays or, if they're available, they may be prohibitively expensive. Any number of misconceptions can lead to assumption errors and increased costs.

A cost model helps you regenerate your budget number as you work to decrease your uncertainty. Assumption errors cost money because they are the embodiment of uncertainty unfunded. Unfunded uncertainties require expensive crisis-led responses to resolve.

CAUTION: Avoid the costly assumption error of expecting staff to both move a data center and simultaneously perform their existing full-time jobs. Budget for a contingency plan to address this at the first sign of trouble.

Budgeting for a data center move should be less about quickly getting to a single number and more about getting the right cost model in place to successfully deal with assumption errors and risks. Said another way, a cost model helps you adjust when underlying assumptions change or are clarified.

More than that, a cost model documents all assumptions, including those changing with time or events. The model illuminates specific budget impacts early in the process to minimize costly surprises during the project's implementation. To determine the effect on costs, be sure to vary your assumptions.

CAUTION: This sensitivity analysis may get skipped in the rush to feature that single cost number on a presentation slide.

Create a Realistic Cost Model

Expect budgets to be imprecise during the early phases of a data center move. But with careful attention to detail, you can produce a realistic and adaptable cost model in which precision improves as you decrease your uncertainty. You'll be ready for cost hazards tumbling down during your journey.

A spreadsheet is a good tool for your cost model as you maintain the *move narrative*, assumptions, and budgeted costs. That way, when numbers change, you will see their effect on your bottom-line number.

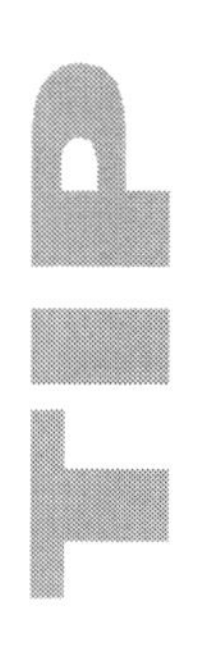

ELEMENTS OF A SUCCESSFUL COST MODEL

Include these components in your cost model:

- *A documented* ***move narrative*** *with specific categories (see category examples below).*
- *A list of your assumptions in narrative form.* ***Don't make a budget all about numbers. Narratives inform the interpretation of those numbers.***
- *Budget elements coupled with the uncertainty gauge. For example, create a column labeled "Uncertainty" using a scale of 1=certain, 2=mostly certain, and 3=uncertain.* ***Uncertainty ranking calibrates your numbers and helps you direct resources to chase uncertainty.***
- *A plan for assessing gaps in your analysis.*

Naturally, there are costs associated with budgeting. Most companies don't bother counting budgeting costs, but they are important when finalizing your budget. Ensure you include these typical tasks:

- *Budget reviews*
- *Recurring budget revisions*
- *Sensitivity analysis of the budget*
- *Training and understanding to know what budget elements are required*
- *Internal or external resources to provide early move plans and subsequent costing.*

Example Cost Model

Creating a cost model is the first step to assembling a budget. You'd run the cost model the first time to get the draft budget. Many professionals stop once they get their first budget number. What are the chances your first budget number comes from a perfectly constructed cost model? Almost zero.

As you look at the example that follows, realize this budget depends on the *move narrative*, the move assumptions, and the ability of the team to execute on those assumptions.

CAUTION: *Your actual budget will be drastically different than this example because the costs used are purposely unrealistic.* That's because using real costs would be counterproductive to understanding that grabbing a cost model here (or online) won't give you the easy button to shortcutting a budget analysis.

Let's examine a simple budget for a small move. For this example, we are moving eight production systems requiring a single move event. The destination will be a leased data center suite and the origin is an owned facility that also needs decommissioning (the proper removal of data center support systems).

From that brief description, these are known elements:

1. Eight systems are moving.
2. The destination is known.
3. The basics of the ***move narrative*** are known and described as one move event.
4. The origin requires decommissioning and there will be duplicate carrying costs.
5. The destination is a leased facility so time is needed to negotiate and sign a contract.

Important details affecting costs are still undiscovered. Let's suppose our governance body has further stipulated these constraints:

- Internal resources will handle the move.
- No downtime can be tolerated for customer-facing websites and e-mail.

What should command your attention immediately? Discovering the detailed knowledge of those eight systems. Knowing you lack specific system knowledge, how will you improve that information gap to remove it from the uncertainty shadows? It appears a cost element is needed for that effort.

Do you know if it's even feasible to have zero downtime for the systems in question? Another cost element is required to properly account for a zero downtime constraint.

CAUTION: Internal moves notoriously ignore the double-counting jeopardy. That's when company leaders forget their internal resources have "real jobs" in addition to the data center move tasks. So carefully note the internal cost elements to accommodate all the hand waving and forgetting that happens.

SPIN A BUDGET WHEEL

Every day, I get requests to disclose budgets for moves I have completed on behalf of clients. The pitch goes something like this: "I just need to see an example to get started. My boss needs a number today."

Well, you might as well spin a budget wheel and give your boss that random number. Budgeting is hard work precisely because it must reflect your ***move narrative,*** *your available resources (or lack of resources), and your likely path of executing your data center move. If a quick pick method like choosing six lottery numbers exists, I'm certain you'd have a similar outcome. You'd lose most of the time.*

Organize by Move Phases

Experience suggests it's best to organize your cost model by move phases. This encourages a systematic approach to find missing elements inside of all move phases. Remember, the purpose of a cost model is to *predict future costs* and *drive out uncertainty* by running the model when new data arrives.

My simplistic cost model for the narrative previously described can be found at this link: http://datacentermoving.com/costmodel.html. In this example, I purposely used incorrect cost number details to help you learn the lesson of cost models. The first number doesn't matter and isn't right anyway. At 2.6, that pesky uncertainty meter is way too high to accurately predict the future. That means don't bet on that first budget number. In fact, you're gambling on having *any* kind of chance to hit it when the uncertainty meter is high.

Uncertainty Meter	Move Phase	Task Details	Hours Estimate	Rate	Subtotal
(1=certain, 2=mostly certain, and 3=uncertain)					
1	Governance				
		Establish Governance	100	75	$7,500.00
		Regular Governance Meetings	100	75	$7,500.00
3	Budgeting				
		Initial cost model preparation			
		Vetting, Presentation, Revision			
1	Site Selection				
		Destination already selected	0	75	$0.00
		Contract Negotiation and Legal Sign-off	100	75	$7,500.00
3	Pre-Move Planning				
		Physical Inventory	100	75	$7,500.00
		Application Inventory	100	75	$7,500.00
		Feasibility of Downtime	100	75	$7,500.00
		First Draft of Move Plan - One Move Event	100	75	$7,500.00
3	Teardown				
3	Transit				
3	Arrival				
3	Re-Assembly				
3	Post-Move				
		Duplicate carrying costs of the orgin - 3 months	3	1000	$3,000.00
		Decommissioning of orgin	100	75	$7,500.00
Average 2.6 Uncertainty		**Subtotal**			**$63,000.00**

Table 1 - Example Budget

This example shows a cost of $63,000 to move these eight pieces of equipment. But the uncertainty number is 2.6 (meaning mostly uncertain). With 7 out of 9 move phases in the first column labeled as *uncertain,* how can this be made less uncertain? Budget for uncertainty.

CAUTION: Strategies exist for budgeting for uncertainty. Contingencies, wild guesses, and soliciting quotes from suppliers are three methods. None of these actually improves the prediction; it just makes you feel better about publishing your number.

Let's take one of the three methods—soliciting quotes—and run a thought experiment to understand the peril.

You get three quotes for moving these eight pieces of equipment. Because your brain is wired to select the lowest bidder, you subsequently tell your manager how much it should cost based on the lowest bid. If your firm will still carry out the move internally, then the time spent for soliciting costs consumed scarce resources. In addition, taking the time to do this has delayed the move execution.

Will you properly account for the time and effort it took to run a solicitation in the cost model? Will the bidder's assumptions and skills carry over to your internal team tasked to hit the budget number? Is there any time remaining on the shot clock to move the equipment after conducting the solicitation? Or does the internal team have to take shortcuts to meet a prescribed deadline?

Clearly, understanding the fundamental phases of the move process seems like a better strategy than guesswork to improve your budget prediction.

Cost Model Categories

I get calls frequently requesting the average cost per square foot or cost per server or cost per application to use for budgeting. Look at the *Preliminary Cost Model Categories* that follow. Do those categories lend themselves to normalizations that can be trusted?

Instead, build your cost model from categories that matter to you. Your firm makes economic decisions based on *your* cost model and will likely find savings in one or more elements. There is no shortcut here; only a systematic effort to drive out uncertainty.

PRELIMINARY COST MODEL CATEGORIES

Listed here are common cost elements for most data center moves. Your unique situation will likely require additional items.

1. *External Connectivity*
   a. Wide Area Data Network
   b. Wide Area Storage Network
   c. Voice
   d. Wireless
2. *Internal Connectivity (Segment by location if multiple sites are involved.)*
   a. Copper
   b. Fiber
   c. Wireless

3. *Move Elements by Phase*
    a. Site Selection
    b. Pre-Move
    c. Teardown
    d. Transit
    e. Arrival
    f. Re-Assembly
    g. Test and Validation
    h. Post-Move
4. *Decommissioning*
5. *Capital Costs (for equipment purchased prior to the move)*
6. *Duplicate Carrying Costs (for items like circuits and facilities)*
7. *Outside Resource Costs*
    a. Move plan creation
    b. Move project management
    c. Equipment vendor re-certification
    d. Data center design experts
    e. Tenant representative
    f. Real estate commissions
    g. Special transportation if needed
8. *Celebration event planning*
9. *Awards, bonuses, and gifts*

These elements form the basis of your line items. They are a beginning, not the finished model.

CAUTION: *If this is your first cost model, seek help from your internal budget analysts. A refined cost model should differentiate between recurring and one-time costs. It's a powerful decision tool and the right resources should be committed to build it.*

Comparing Costs to Budget—a Paradox

After every move, I review the original budget number to compare it to the actual cost. Would it surprise you to learn none of the budgets perfectly predicted the future? Welcome to the budgeting paradox of data center moves.

But why does this keep happening? Think of it this way. In the beginning, you knew very little about the relocation task in front of you; your budget reflects that uncertainty. If you fall prey to expending extraordinary effort on planning, your technical and political environment changes because this planning effort takes a long time. The result is an elongated schedule, more costs, and more resources dedicated to crafting the perfect budget—even while it rapidly accelerates into obsolescence!

CAUTION: If you do very little planning, you will spend top dollar in a crisis mode fixing the problems you created during the relocation, perhaps in a highly visible and damaging way to your firm.

Each data center move comes with this budget challenge. Accept that timely budget decisions must be made with imperfect data. That's your reality. Find a balance and review your cost model regularly to anticipate any contingencies.

Next, let's look at the site selection phase.

SITE SELECTION

WHY ARE YOU MOVING YOUR DATA CENTER?

Most moves happen for economic reasons. Thus, your site selection may be restricted by cost considerations. However, it's important to understand the executives' motivations behind reducing data center costs.

Are they looking for the lowest per square foot cost? Are they asking for the best operating costs over the life of the data center including labor costs? Is a move of both headquarters and data center at the same time an efficient cost saver?

Uncovering these expectations and their constraints benefits your site selection process. Why? Because the strategic reasons for a site selection have consequences.

There's a famous example of a rental-car corporate relocation being made because the CEO (since fired) wanted it close to his home instead of optimizing the location to benefit the business. Make sure you understand the strategies and constraints involved so these types of blunders aren't repeated.

Don't Expect Magic Answers

Give the site selection process plenty of attention and time. Choosing the best site for your organization means meeting all the requirements, not simply bending to harsh economics. Expect others to rigorously challenge all site selections and prepare accordingly. This means document the decisions so they stand up to reviews by other experts. Some companies employ one team to research sites and another team to evaluate the results.

Site selection happens infrequently for most firms. Do you have the discipline to form the right team and get educated about a successful process? If not, seek a qualified firm that will confidentially review and evaluate your sites.

Some site selection teams experience immense pressure to reduce costs quickly. In these circumstances, don't let that pressure obscure hidden costs that will surface after the move. Spotting hazards under these conditions is difficult due to the time pressures, but by avoiding blunders down the road, your discipline will pay off.

Is hiring a tenant representative right for you? Commercial real estate evaluations should include tenant improvement allowances, signage considerations, and many important contractual details affecting your choice. Using a tenant representative to narrow your choices is an option to expertly guide you in unfamiliar markets. They can answer concerns quickly without the elongated research cycle you'd experience going it alone. This allows you to stay focused on the task ahead.

Remaining anonymous during site selection is a common concern a representative may address with you. In real estate, you won't find local insights in databases; they exist in dealing with professionals who have deep experiences and important relationships. Another common (and costly) mistake is failing to consider a long-term lease to help you lock in important tenant improvement allowances.

CAUTION: Concessions can be substantial. Hiring an experienced representative can help.

Building Your Own Data Center

Many organizations build their own data centers but few consider the total cost of ownership. Instead, the typical winning design is the lowest initial capital cost, *not the lowest total cost of ownership over the life of the facility.*

Further, this process practically ensures the firm repeats the same mistakes of the past 30 years. That's because poor compromises are justified in the name of lowest initial capital costs. The resultant data center tends to be the opposite of an efficient room. It can cost more to operate over the lifetime of the facility, and you risk ignoring critical technological advances in data center design.

Cut corners at your own risk

Exhaustively listing data center design mistakes is beyond the scope of this guide. But know this: Many poor choices result from the same fundamental issue—that is, *not enough capital and time to design and build a proper data center from scratch.*

As a result, engineering shortcuts to cut costs will be required, yet budget-driven building tactics can create cost burdens over the long haul. Decisions made about the build dramatically affect the total cost of operation before the first server is ever activated. Commissioning your new data center, for example, should never be skipped.

What does it mean to commission a data center? The major operational components of a data center are power, cooling, fire and life safety systems, and management systems. These systems work together to provide a 24/7 data center that supports critical business applications. Commissioning is the disciplined approach for confirming that *all* of the major systems function together to support the 24/7 mission.

CAUTION: Commissioning not only applies to newly constructed data centers; any time you inherit a data center, consider some form of commissioning. That way, you can avoid operational surprises and unexpected downtime.

Difference between "green field" and "brown field"

The industry term for your new data center is "green field," which means the facility is new, engineering and start-up data are readily available, and installation engineers are accessible. A "brown field" refers to an existing facility and presents significantly more challenges for commissioning. For example, important records might be missing or outdated; equipment may be improperly maintained and serviced. Yet these "brown fields" must still support a 24/7 data center operation. They deserve attention to the commissioning process before you occupy the new data center.

Schedule compression occurs when the end dates don't move as more work is shoved into a shorter timeframe. It is a harsh master forcing trade-offs along the entire path of a data center move. But shortcuts taken during commissioning will lead to expensive surprises when the entire data center is supporting business-critical applications.

Here's an example of what can happen with out-of-sequence testing.

I reviewed the commissioning certification for a newly built data center and found the Emergency Power Off (EPO) had been tested out of sequence. Specifically, it had been tested prior to the finished installation of all of the supporting systems.

Naturally, I wanted to push "the big red button" and retest the conclusions of the report. What did I find? Two things:

- A misconfiguration of an unrelated door system caused minor perimeter alarms for distant doors.

- A misconfigured fire suppression system auto-dialed the fire department even though no fire signal was present.

It's best to correct minor issues such as these before moving into your data center.

WHERE TO FIND COMMISSIONING GUIDELINES

The American Society of Heating, Refrigerating, and Air-Conditioning Engineers (ASHRAE) publishes commissioning guidelines such as ***Guideline 0—The Commissioning Process*** *referenced by most practitioners. Equipment manufacturers also have documented commissioning procedures specific to their gear.*

It can take months to formulate the startup sequences and testing schedules of all the systems in scope. If your data center is being built from scratch, ensure your schedule has the appropriate resources and time devoted to commissioning.

Considerations for Data Center Build

Did you know that airside cooling is the least efficient way to cool computer equipment? Yet, it's the leading method still chosen for cooling in spite of the science that says liquid removes heat more efficiently than air. Let's assume you're not considering liquid or even immersion cooling techniques because of their high initial capital requirements. Lowest-cost choices override better choices and prevent you from realizing substantial savings over the life of the data center.

What can be done about airside cooling systems if your capital is inadequate for other solutions? For starters, eliminate the cost of the raised floor and place your equipment on properly prepared concrete slab.

Pushing air down into the vast under-floor plenum to be pushed back up through perforated tiles requires two things: Expensive computer room air conditioners (CRAC units) and a lot of power for their fans. Eliminating the electrical power for these fans represents a 75% savings in power loads for air conditioning. Not having a raised floor represents a substantial capital equipment savings. It also saves the labor, cost, and time required not to build it.

Using a raised floor would be an expensive mistake resulting in higher total cost of ownership, specialized CRAC units, oversized rooms with more capacity than will be used, awkward perimeters to accommodate the CRAC units, and longer construction timelines.

Instead, put your redundant cooling systems on the roof and let the cold air fall as the heated air rises. This means designing your data center to use cold aisle containment systems *from the start* instead of as an afterthought.

The use of cold aisle containment means you use less expensive but equally capable rooftop or ceiling cooling units. Instead of fighting physics with increased power usage for fans to maintain static air pressure in a wasteful, raised-floor plenum, you accrue cost savings by letting cold air fall and hot air rise without mixing.

When considering building from scratch, honestly ask these questions:

- Could your new project measure up to an existing purpose-built site?

- At what capital cost would the build require?

- What operating expenses do you need to keep it running?

- How many of these criteria are important to your firm?
    - Survive tornados and other natural disasters
    - SSAE16 SOC-1 Type II Certified

- GSA approved
- Meets HIPPA requirements
- Biometric access and audit logging with two-factor entry
- Redundant bandwidth
- Competing carriers for bandwidth price reductions
- Diverse inbound carrier conduit routes
- Prioritized for generator fuel delivery
- Power redundancy—two or more utility feeds
- Backup power redundancy—multiple generators
- Roof leak protection
- Air conditioning on generator power
- Redundant infrastructure of N+1 systems

CAUTION: If you consider a build of your own data center, be precise about all the costs and the requirements you can't meet when presenting the alternatives for a decision. Tell the truth about costs and the trade-offs made to meet those costs. Your business deserves you make a well-informed decision when weighing your options.

TURNING LIES INTO TRUTHS

Personnel in a firm I know misrepresented that their facility department could build a data center despite a lack of data center experience. It would save the firm money compared with the costs of a purpose-built data center from specialized experts.

What happened? Those in charge took short-cuts to make the lie the truth. They created a design to show cost savings instead of a modern, efficient data center. My review found these issues: lack of cold aisle contain-ment, inadequate roof designed for an office building and not a data center, lack of sub-metering to track electrical usage, and undersized power infrastructure to hold down costs. This compromised the data center's efficiency and added operational costs.

They successfully demonstrated they could build a cheaper data center, but the result was a data center that would be costly to operate with poor compromises made to hold down initial costs.

Alternatives to Building Your Own Data Center

Many choices exist and more appear every day. At the highest level, you will find variations of these options:

- Colocation—You rent locked cabinets beside other customers in a data center.
- Caged Colocation—You rent a locked caged area of your required dimensions and build out the footprint to your specifications.
- Suite Colocation—You rent a locked suite and build out the required footprint.
- Wholesale Colocation—You contract for very large footprints and build out the required footprint within a self-contained space excluding other tenants.
- Modules—You buy or rent modular containers and place them on your property or inside a colocation facility.
- Cloud—You move your applications, not your physical infrastructure, to a service provider (or several providers) under an agreement specifying service levels for availability. (A later section addresses common concerns with a Cloud-based move.)

Choices are influenced by many factors, but two are quite interesting: 1) underground facilities that don't suffer from most natural disaster concerns and 2) container modules that speed time to build while promising to scale to demand.

CAUTION: Give modules and underground facilities serious consideration to determine if they fit your needs.

What are some of the cost model categories specific to site selection?

- Assemble a site selection team.
- Pay legal fees for site selection options contracts.
- Allow travel costs for site visits.
- Hire expert advice for power and connectivity options for sites.
- Study "build" versus "buy" trade-off options.

If the options available are more numerous than you realized, then you know why a disciplined approach to site selection is critical to your success.

Next, let's take a close look at tasks you and your team have to complete before your data center move.

PRE-MOVE PLANNING

WHAT NEEDS TO BE COMPLETED BEFORE YOUR DATA CENTER MOVE?

The Pre-Move Planning phase contains all the activities to complete before a single piece of data center equipment is moved. The consequences of rushing through Pre-Move Planning include unexpected hazards, non-optional costs, and confused staff. Like a three-alarm blaze on a cold, wintery night, all of these are likely to show up at the most inconvenient times.

Taking on an inherited schedule and non-optional milestones means you'll encounter headwinds on your journey to a successful move. Many of my moves started the same way, so I understand the panic and the overwhelming sense of impossibility you may feel.

CAUTION: Many strategies you've been taught about a project plan may doom you before you start.

Practical Methods

To help you conquer your fear, let me define and explain the components of a *move playbook* as a container that comes with a recipe for creating your unique playbook. I also provide templates for Stakeholder Communications and Go/No-Go Criteria for use in your new playbook.

WBS

A Work Breakdown Structure (WBS) is the most familiar tool to project managers. I'm seeding your project with three high-level WBS examples; you provide the detail. Note: These are examples to spur your thinking; they're not meant to be a boilerplate to copy.

Rehearsal

Rehearsing a data center move is the single best thing you can do to validate your plan. I explain the concept and help you understand how to organize your rehearsal for success.

Cloud

Although moving to the Cloud is *not* magic, it *is* different. I explain differences and give you advice to tackle a move to the Cloud. Remember, the same professional discipline is required for a Cloud move as a physical move. Don't let anyone's marketing hyperbole trick you into taking shortcuts for a Cloud move.

SOWs and Punch List

Most firms have a procurement department responsible for writing Statements of Work (SOWs). All the pointers in this guide will help your procurement people write a SOW tailored for your data center move. This pre-move section also introduces the concept of a Punch List and provides a brief review of what you've learned.

Roadmap To Your Move

Complex moves are broken into several discrete move events, each one requiring its own pre-move planning. You will have your own approach to this task, but it helps to have some structure in mind for your first attempt. Segmenting the activities by origin and destination will help initially.

Eventually, the complexity may overtake you and additional approaches will be required. You are building a roadmap to your data center move in this step. This becomes the coveted boilerplate document you may have spent days searching for online to avoid the work.

I call this document the data center *move playbook.* Because each move has unique requirements, I build a playbook of these from scratch for every client. I've also reviewed many data center relocation plans. Some are simplistic and some are complex, but most suffer from a single and fatal flaw—that is, *everything has to work correctly in the prescribed order.* That means everything has to go right.

Is that a likely event in your world? Answer that question in the context of an airline flight. Suppose there's only one linear sequence of events that results in a successful landing of your plane. As long as nothing along the event timeline misfires, the landing will be just fine. But are you ready to take off knowing everything has to go right in a precise order to successfully land? That's right. No deviations, no unexpected issues, and only one solution for landing. All other solutions are untested and unknown by the pilot.

Yet, data center moving project managers build a plan that requires everything having to go *right*. Then they expend extraordinary effort tracking in minute detail everything that goes *wrong*. Doing this can consume a disproportionate share of resources and costs. Like landing an airplane, the wrong time to ask about alternatives is during the final approach. As a leader, recognize this planning flaw and counteract it *before* takeoff.

Is there a better way? It might depend on your built-in biases for using or refusing outside help. It might also depend on your willingness to recognize the confirmation bias that sabotages most moves led internally. It certainly depends on the schedule pressure you're experiencing.

The Tyranny of Your Plan

At issue is the method you use to create a complex plan. This method is its own form of confirmation bias and the resultant plan holds a tyranny over the project. To illustrate this point, consider how project managers are taught to create plans. In this order, they:

- Break the work into smaller tasks with start and stop dates. This is known as the work breakdown structure or WBS.
- Sum up the resulting WBS pieces into the project schedule plan.
- Track progress against the resulting plan.

What kind of a plan is this? In reality, it's not a plan at all. When you sum up a schedule based on the pieces of the work breakdown, you've created a guess about the future. You have a hypothesis of how the plan might be completed—one that highly depends on your initial arrangement of the WBS elements. And face it. Your guess isn't based on deep experience with data center moves. You risk trouble starting almost immediately.

What happens when reality doesn't match your hypothesis? In most cases, you do two things: 1) You tell offenders to try harder to do exactly what the plan says, and/or 2) You plan again, attempting perfection while expending more effort.

CAUTION: These two things consume scarce resources while doing nothing to improve the chances of success.

Do you see how a plan can hold a tyranny over actually completing important work? Instead of recognizing that your guess about the future has been proven wrong in the face of reality, you expend extraordinary effort to tame the turbulence with a cycle of blame and re-planning.

TIP

EMPIRE STATE BUILDING SUCCESS FACTORS

Consider the story of how the Empire State Building was constructed in the era well before computers and WBS. These key success factors for the 18-month construction of the iconic New York City skyscraper can be applied to data center moves:

- *Extremely experienced builders are used. The building's tenants don't design it; the experts do.*
- *Constraints drive schedule. The concept of time boxing (what can be done in this time period) versus scope boxing (how long will this scope take) is used.*
- *Break dependencies instead of tracking them. Scheduling dependencies takes far more effort than breaking them to keep other steps unaffected. For example, while site selection is a necessary step, don't let that step prevent a deep dive into the*

applications you are moving. Instead, determine how each element of work can make progress without waiting for a previous element to finish completely.

Typically, weak data center move plans are created by those with built-in confirmation bias and no deep experience doing the move itself. Don't be surprised when those plans hold a tyranny over your project.

Your Plan Is Still Just a Guess!

The reality is that most WBS-built plans, while impressively detailed, still guess about the future. Yet a guess that's deemed "incorrect" triggers a massive increase in measurement activities to make the plan right. After that, the resources used for measurement are not being used to solve the problems—a wasted opportunity.

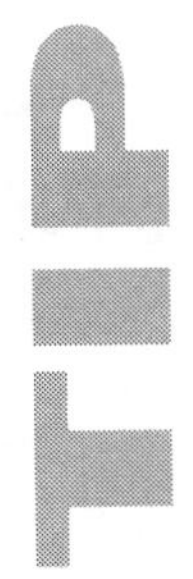

PROJECT MANAGER—HELP OR HINDRANCE?

A merchant who operated a market at the top of a hill relied on vendors to supply the market. They had to pull their wagons up the incline several times a day. Wanting to increase sales by decreasing the time for wagon trips, she hired a project manager to address the problem. Afterward, the trips took even longer! When she asked a vendor to explain it, he said,

> *"I could pull it faster if you can get that project manager out of my wagon."*
>
> *There's a lesson here. Practitioners may view project managers as a hindrance if the effort to support them exceeds the effort to complete the task. Project managers may view practitioners as uncooperative because of their singular focus on task completion over status reporting. Balance and communication are required throughout the data center move.*
>
> *Meet halfway. Learn about the data center move process before trying to project manage it.*

Why burden the people responsible for the work by loading their wagon with project managers tracking their hypothesis about the future—one that's already proven wrong? Why expend resources on the wrong effort? Yet, many move plans continue to consume scarce resources at alarming and accelerating rates—all the while tracking their trajectory into the ditch without correcting course or understanding the effect of excessive tracking. Predictable outcomes include more meetings and less progress.

I suggest that plans based on experience are reliable compared with plans based on wishful thinking (such as the intricate guesses generated by WBS-based methodology). Given that, you may not have patience to learn the process of building plans based on experience. You simply want a sample plan right now.

If you believe this guide represents a cargo hold of "easy" buttons, you'll find the rest of this section frustrating. But if you want to learn about the process of creating a data center *move playbook* you can use, keep reading. Here's what you can expect:

- How to outline ***move playbook*** elements
- How to conduct a data center move rehearsal to exercise your completed playbook
- Awareness of confirmation bias and how it influences many aspects of human behavior including ***move playbook*** construction. For example, when the technical staff primarily writes the playbook, it is weighted to familiar strengths such as technical diagrams and technical recipes. When project managers write them, they set up milestones and work breakdown structures (WBS).

My own bias is weighted with construction-style punch lists containing clear owners of tasks using possession arrows (explained at the end of this section).

CAUTION: Avoid constructing a playbook under the streetlight of familiarity. Instead, ensure the entire landscape of your move is well lit with understanding. This balance will help you achieve a comprehensive playbook for your move.

Move Playbook Explained

Before I began my consulting career, I spent more than a decade with IBM inside many large data centers. Typically, the data center operated in a state of constant change with equipment moving in and out frequently. In fact, this equipment was much larger than the hundreds of smaller servers you find in today's data center. Moving a data center today is challenging for many reasons. A few of the challenges include an abundant mix of dissimilar vendor equipment, a constantly changing workforce responsible for the applications, and the speed to change that the business demands.

Rarely is there an accurate physical and logical inventory of the contents of a data center for a variety of reasons. However, two elements are critical to your decisions about moving the data center: 1) How well do you understand your applications? 2) What downtime will your business tolerate?

Beyond that, these seven elements make their appearance in every *move playbook* I build:

1. Master Inventory
2. Application Dependencies
3. Master Timeline
4. Sequencing or Batting Order
5. Stakeholder Communications Matrix
6. Contingencies
7. Go/No-Go Criteria

Let's define each of these seven *move playbook* elements before exploring how they are developed.

Move Playbook Definition of Elements

1. Master Inventory

This is a complex inventory with information about hardware and applications across all geographies necessary. Columns contain a rich source of additional move-related information that many move activities will parse.

2. Application Dependencies

This document is crucial to a successful move. It contains information from the perspective of the applications. Because a data center move is about the successful relocation of the business applications, it only makes sense to spend time understanding all the application dependencies.

3. Master Timeline

This is a favored document of project managers and executives alike. However, it's only a prediction of the future. The Master Timeline should frequently be briefed at a high level. It helps calibrate everyone's understanding of the future prediction contrasted against actual events.

4. Sequencing or Batting Order

Planning the precise shutdown, teardown, re-assembly, and startup is important. It's even more important to rehearse this batting order. Your practice will reveal the need for important adjustments to the move sequencing.

5. Stakeholder Communications Matrix

This matrix declares, in advance, the stakeholders who will receive specific messages during your move. Using this template-based approach will detail the frequency for messages sent through the available channels. An unexpected benefit of building this matrix is its exposure to the stakeholders who influence important move-related tactics.

6. Contingencies

Be sure to document how to handle elements of your plan that may fail. Unless you are a rare professional who can build a plan in which everything goes perfectly, don't skip this section. Rehearsals will expose areas where a contingency plan is needed and where you document rollback and recovery scenarios.

7. Go/No-Go Criteria

Every move event approaches the moment of truth when you decide to go forward or fall back. Specifying your decision criteria ahead of time gives stakeholders an opportunity to approve your thinking outside the pressure of a crisis.

Your specific move will certainly have additional sections in your playbook and that's fine. It's the *process* of building and rehearsing the playbook that gets your entire staff singing from the same sheet of music as you execute your data center move.

CAUTION: Know that building a *move playbook* is not a linear sequence of events. The resources you need for each component varies. Your inherited schedule and limited budget will tempt you to cut corners and rush the playbook development. Rehearsals are important and should not be skipped.

What Happens in a Perfect World

In a perfectly executed move, you'd approach it sequentially, and you'd only do each task once with perfection. It might look something like this:

1. Gather the Master Inventory.
2. Gather the Application Inventory.
3. Decide how many move events will be required.
4. Build the Master Timeline.
5. Sequence the Batting Order.
6. Build the Stakeholder Communications Matrix.
7. Rehearse the playbook to reveal the contingencies required.
8. Decide the Go/No-Go Criteria.

Now examine this linear process. Which items can be executed in parallel? How many items require revision and tuning as you learn more? Is there a better way? Perhaps.

Quite a bit depends on your internal culture and what I call the *organizational coefficient of drag*. This is the amount of energy required to gain momentum on a new project within your firm. If your data center *move playbook* construction must follow all established rules for a new project, then it's likely your organizational coefficient of drag will extract a high price in terms of time and resources to even get started.

That's not to say formal project-management procedures such as project charters and methodologies aren't important. If you are lucky, you'll find a champion to *accelerate* through those formal requirements. However, most internally led moves don't accelerate; they get bogged down right at the beginning. Before any meaningful work on the *move playbook* construction begins, aim to overcome the organizational coefficient of drag.

CAUTION: Address this issue immediately, speaking truth to power if necessary, to avoid this common startup issue.

FACE THE REALITY OF YOUR CULTURE

Remember, a data center move operates in the cultural and procedural environment that already exists in your company. Functional or dysfunctional, face the reality of your culture and enlist allies for assistance.

If you overlook the organizational coefficient of drag for now, consider what tasks can be done in parallel. Perhaps these four:

- Gather the Master Inventory.
- Gather the Application Inventory.

- Build the Master Timeline.
- Build the Stakeholder Communications Matrix.

Recognize these tasks will finish at different times with varying accuracy of results. After all, *move playbooks* are multi-author efforts in large firms. This means your attention to specifying a container for the work will help those writers focus on content. Be sure to revise your container as new information is uncovered.

Even before any of these tasks finish, involve your governance team to brief them on your approach. This is a great time to reinforce the need to support the rehearsal of the playbook. Once your first draft is complete, add your Go/No-Go Criteria and also publish the draft for vetting and additional revisions.

The container that follows is a suggested way to start each of the playbook sections, starting with the Master Inventory.

Seeding Your ***Move Playbook***

1. Gather the Master Inventory

Often, the initial physical inventory can be assembled with automated tools you have available. Supplement the basic information about the inventory with data to use in later stages of your move. The extra data might include:

- Value—Equipment value helps you estimate the additional insurance to procure prior to moving. Supplement with insurance sold by replacement value of your equipment.

- Support—Knowing what equipment is still under support is important because some equipment might not be worth moving. Costs to move unsupported systems may exceed buying new equipment and should be contrasted against alternatives.

- Operating System (OS)—Knowing all the versions of the operating systems will influence your move strategy.

- Power supplies and their specific power connectors. Note any unique power connectors or type of power required to avoid power surprises at the destination. A thermodynamic equation turns power into the cooling you need. Use your power data as an important check when sizing cooling systems.

I have observed equipment inventories with more than 100 columns of individual descriptions. Some of these descriptors come in handy; some are simply noise; some prove their value on every move. These include:

- IP addresses—List both existing and newly assigned IP addresses.

- Technical Owner—Have the system administrator responsible for the equipment available during debug.
- Business Owner—List the business group most affected by potential downtime.
- Move Type—Designate a type—physical, logical, retirement, or no move—in this column.
- Membership—Indicate if the equipment is a member of the production environment or the development or test environments.
- Backup Validated—Note if the backups have been restored and tested. It's worth repeating that, if you haven't done a restore test, your backup is completely invalidated.

Later in your planning, you might consider these three additional columns to help manage your move sequencing:

- Origin Rack Location—Note the physical location of the equipment prior to the move.
- Destination Rack Location—Note the physical location of the equipment after the move.
- Move Sequence Number—Indicate which move will trigger the move of this equipment.

Many think of the Master Inventory as simply a spreadsheet. However, non-spreadsheet components should also be collected as part of your inventory including:

- Elevation diagrams of the equipment location inside their cabinets

- Network diagrams showing the interconnection

- Digital photos documenting unique equipment or complicated interconnections

CAUTION: Don't forget to inventory your circuits because many of these require special handling. Contractual requirements may require written notices and specific termination clauses. You can leverage your need for new circuits at the destination to minimize the termination costs at the origin. Think in terms of 90 days or longer to handle circuit terminations. Circuits are a long lead-time item for your move. New circuits may need even longer timelines if facility construction is required.

Finally, do an origin site inspection to validate your inventory collection. Are the machines in the correct locations? Is there special equipment present that was missed? Are the observed interconnections more complex than the simple diagrams provided? Always open the back of the cabinets to get a sense of the labor required to disassemble and the important role proper labeling will play.

2. Gather the Application Inventory

Application inventories are the most important aspect of your data center move. Your business depends on applications. Explicitly record the downtime allowance of those applications to inform your move planning about critical priorities.

After decades of moving applications across data centers, I've developed this method of segmenting them for the purposes of moving. Here are the steps:

- ***Membership***—Designate if the application is customer facing or internal. This helps you optimize for downtime, stakeholder communications, and magnitude of resources required. Further segmentation is helpful to identify applications moving at the same time. Understanding this application affinity helps you optimize your move.

- ***Type***—Classify the application as a web server, file transfer server, database server, mail server, application server, or file server. This helps you organize the type of skill required for each of the applications.

- ***Security***—Note the specific security requirements to include single or two-factor authentication, firewall security zone, regulatory compliance, and internal policy requirements. This informs your resource planning because the process of moving applications often exposes new security concerns. Extra care is required to certify your security requirements survived the move.

- ***Dependencies***—This broad category can become a black hole consuming extraordinary resources. Many applications depend on a myriad of other services. Databases, file servers, and external connectivity are three examples. List what you know. Complicated applications will require deeper dives by experts to uncover all the dependencies.

In virtually every move, I encounter applications in which no domain knowledge remains in the organization. What do you do when confronted with moving a mystery application? The technique I use is to ask questions designed to gauge the amount of acceptable risk that can be tolerated against the cost of removing that risk.

Welcome to the triage portion of your move. Here are four sample questions to get you started:

1. Is the application customer facing or internal? If customer facing, is there budget to hire a subject matter expert for moving the application? If internal, is the application critical to driving revenue?

2. Are you still paying for support? Then consult the vendor for help covered under your support contract.

3. Can the application be virtualized for the purposes of testing? Often, the original application can be left untouched in the origin while a virtualized surrogate application can be tested in the destination. Through trial and error, the risks can be squeezed out until the application is ready to move. Plan for trial and error consuming more time as the trade-off.

4. What alternatives exist to replace the application? Because many mystery applications are legacy applications, there is likely a modern alternative that can replace the legacy application completely. Recognize this isn't a shortcut, as migrating to a brand-new replacement carries additional risks on top of a data center migration. However, your firm may be better off in the long run by considering the alternatives before investing in moving the legacy application.

Many methods are required to paint an accurate picture of applications and their dependencies. There are automated tools to help map application dependencies and visualize them for understanding. Use them if they fit your budget.

Don't wait for completion of the entire application inventory before acting on the data.

Is it part of a larger technical ecosystem? Who are the end users and what downtime can be tolerated? Waiting for a comprehensive application mapping robs you of urgency and obscures previously fresh details. De-clutter your view instead by acting immediately to decide an application's fate instead of waiting for the entire report.

An expert should log into every application to inspect and validate the dependencies. Application discovery is a very detailed endeavor. I've compiled a beginner's guide to aid your application discovery found at: http://datacentermoving.com/appdiscovery.html.

If your staff is infinitely familiar with the applications running your business, then this discovery step will proceed efficiently. Budget enough time when there are gaps in the internal knowledge requiring deeper inspection.

The most time-consuming aspect of your data center move may be the application inventory and subsequent application dependency mapping and affinity grouping. When I review move plans, this task is usually underfunded. It's unreasonable to expect that staff with custodial responsibilities will also have the extensive knowledge of every application inside the data center.

It's also true that vendors will struggle with this task for the same reason—multi-vendor environments running multi-platform applications are complicated. Some may have hundreds or more configuration options. Some will no longer be under a maintenance contract. It's a vexing environment for executives because costs increase rapidly before clarity emerges when producing a plan that can be executed.

A companion guide that details every application you might encounter with tips to move them would be very helpful. That mythical guide is the ocean of applications, so I can only warn you that this application inventory phase can be quite complex.

Like unregistered vehicles on a roadway, applications spring to life with no registration requirements when they are installed. They operate for years or even decades and little information is consciously kept about their owners, their users, the true dependencies they have, or the resources they consume. Some may operate like empty busses on inefficient routes until their uselessness is discovered.

Our industry seems to be missing an entire discipline—like fleet management for applications—from birth to death. Applications could be mobile-ready if they were actively managed from the moment of their birth. Instead, their mobility is a once-in-a-career event to be managed as part of a data center move involving time-consuming expertise to uncover their dependencies.

Be vigilant during the application inventory phase of your move. It's likely to be the most frustrating and expensive part. Again, use automated application mapping tools if your budget permits.

CAUTION: Application discovery is expensive, but application ignorance is fatal to your data center move.

3. Build the Master Timeline

Building the Master Timeline is not a one-time activity. For it to have value, it must be accurate. For accuracy, update constantly as new information renders old assumptions incorrect. I could warn you against jumping to Microsoft Project for your first iteration of your timeline, but I'm realistic about how project managers work with these timelines. Instead, I will explain how I build a timeline and maintain it using a simple spreadsheet. You are free to complicate the process with a Gantt chart, and the immediate expectation that your published timeline is the perfect prediction of the future.

Why do I use a simple spreadsheet for the Master Timeline? Because of these properties:

- Simple to publish and view—A read-only version can live in a central location and office productivity software is common meaning virtually anyone can view the file.

- Simple to update—Practically any device can be used to update a spreadsheet.

- Countdown clocks—Use countdown clocks to remind viewers how many days remain before a milestone. Urgency is the currency of progress.

- Versioning is easy—Turn on track changes and you're done.

- Built for repurposing—A spreadsheet is easily imported (over and over again as data changes) into many other tools including Microsoft Project. By using a spreadsheet as the underlying "truth" document, your agility is improved dramatically moving nimbly between other tools. Agility is critical to overcome your organizational coefficient of drag.

What are the steps I use to build a Master Timeline? There is no magic method for this:

1. ***List what you know***—Do you have inherited dates and non-optional milestones? List those first. Designate their source so you know where to go to challenge or confirm the date. So many moves have unchallenged dates causing massive amounts of work because those dates are simply guesses.

2. ***Challenge the dates***—You can't defend dates to your team that remain unchallenged. Find out for yourself what makes the dates legitimate or suspect.

3. ***Review your cost model***—Cost models have many embedded assumptions presuming certain dates will be met. Find and document them.

4. ***Publish your timeline weekly***—The entire team benefits from the most current timeline. Gathering their feedback to validate dates and assumptions is necessary.

5. ***Don't lie to yourself***—Your timeline should not reflect an optimistic prediction of a future when everything goes correctly. When dates slip, update the timeline. How can you change course if your timeline is a lie? Resist the pressure to publish illusions. Publish truth instead.

Focus on the events that matter most and resist the urge to publish every date. Remember, time boxing and dependency busting built the Empire State Building in 18 months. Many agile methodologies have rediscovered time boxing as a very effective technique for complex endeavors.

4. Sequence the Batting Order

How do I build the sequence of events to shutdown, teardown, re-assemble, startup, and test? It may seem counter-intuitive, but I tackle the hardest, most dependent application first when drafting the Batting Order. Because it will touch many other systems, the hardest element will reveal the scope of work remaining.

I depend heavily on rehearsals (discussed later) to improve the Batting Order. Keep these methods in mind as you conduct your data center move rehearsals:

1. ***Categorize***—As you rehearse, note which systems are complicated and which seem to slip

in place. Rework their Batting Order to squeeze out risks by tackling simpler items early in the shutdown phase.

2. ***Consider***—Why did you choose the order you did? What has changed since then to reconsider? Always solve for risk.

3. ***Copy***—Some systems are very similar and their sequences can be copied with minor changes. Take advantage of similarity, but question where divergence requires your diligence.

4. ***Curate***—Notice important dates from a wide collection of documents, meetings, and conversations as your move progresses.

The following template for a Batting Order can be modified for your needs easily:

- ***ID***—The sequential number of the task. Numbering adds a precision in communications during the teardown.

- ***Time***—The expected time when the task will be executed.

- ***Activity***—The explicit activity required. Often, there is a complicated technical recipe referenced in the activity field.

- ***Owner***—The resource responsible for performing the activity.

- ***Comments***—A column flagging any important or unusual notes.

The *Activity* field might look like this for a physical move:

- Assemble all move participants on the dial-in conference bridge to verify everyone is present and ready.

- Shut down machines according to the Batting Order by hostname (host1, host2, and so on).

- Package machines for shipment.

- Verify the manifest and release machines to transportation carrier.

- Transport to destination.

- Arrive at destination.

- Verify move manifest during unloading.

- Reassemble machines in data center and do the racking.

- Sequence startup of machines by hostname (host1, host2, and so on).
- Validate testing of functioning machines by hostname.
- Communicate progress checkpoint to stakeholders.
- Provide final notice to stakeholders with any post-move items noted.

Review this simple example of a Batting Order in a spreadsheet at http://datacentermoving.com/battingorder.html.

How many successful professional teams take the field without learning and practicing their playbook? Your Batting Order benefits immensely from practice as activities are refined and new information added.

Assembling your Batting Order, you may realize you need more than one move event to complete all of the work. Rehearsals also inform the number of move events required. To approximate the number of move events needed, start with three move events. Then test that assumption as your Batting Order takes shape. Do you have too much or too little work in each move event? Adjust to balance your plan accordingly.

Finally, consider carefully the validation and testing required to verify the application performs correctly. It's not unusual for technical staff to have no actual application knowledge to verify correct operations. You might enlist the business users to test their application's functions.

I remember one move for a healthcare provider when a dental application slipped through validation testing. The technology staff could only verify the application services started. They had no credentials to log in to test the database access to x-rays and other patient information. Rousting a dentist familiar with the application, they waited for her testing to end.

"We have teeth!" she exclaimed retrieving a patient x-ray.

In this way, embarrassing downtime was avoided and a valuable lesson learned. Have on hand all the credentials for all the applications *with* corresponding qualified users to validate the application. Make sure it functions correctly after the restart. This checklist should be part of your Batting Order.

5. Stakeholder Communications Matrix

Using a template gives a structure for your important move-related messaging. A basic template documents when messages should be sent, which recipients need those messages, what message channel is used, and who owns the message. Pre-loading the template with message content fosters precision in communications and avoids blunders that reflect badly on the staff for poor messaging. Tagging each discrete message with a Message ID streamlines the process of authorizing these messages via Message ID instead of specifying the entire message body.

A basic communications template includes this:

- ***Message ID***—is the short identifying text for the message
- ***Trigger***—specifies the triggering criteria for sending the message
- ***Recipients***—specifies the stakeholders by name or group
- ***Channel***—specifies the message channel such as e-mail, phone, text, or radio
- ***Message Owner***—specifies who owns the message content
- ***Message Content***—specifies the text of the message

Communicate your move dates to your customers and suppliers. Don't forget to *over communicate* internally so you can address concerns early and less expensively than when in crisis mode. Your firm likely has a department devoted to communications. Enlist these professionals early in this portion of your data center move. Having clear internal communications *prior* to a move creates a shared sense of urgency across the entire firm.

CAUTION: Because communication is key to success, you can't overdo it. Failing to communicate leads to disastrous results.

Go/No-Go Criteria Example

The purpose of the Go/No-Go checklist is to define the criteria for allowing a move to proceed. The example below illustrates the type of elements to include:

- Has the move rehearsal been completed and exceptions incorporated into the Batting Order?
- Have all pre-move communications been sent with exceptions noted?
- Are all resources identified, available, and notified?
- Have validation tests been specified with owners identified?
- Has a contingency plan been discussed with contingency resources on standby?
- Has the Move Commander been identified?
- Has the bridge dial-in and other collaboration tools been tested?
- Has the roll back plan been discussed?

Another way to view the Go/No-Go Criteria is to brainstorm all the ways that might stop your data center move. Most moves are halted at the last minute out of fear instead of legitimate issues. Work through all the fear-based reasons *before* your move to avoid ignorance driving up costs for do-overs.

Next, let's examine a few samples of Work Breakdown Structures (WBS) to further your planning efforts.

PRACTICAL WORK BREAKDOWN STRUCTURE (WBS)

The work breakdown structure (WBS) is a staple tool of many project managers. And while a detailed WBS doesn't ensure a successful data center move, seeing WBS examples can help you jumpstart your move.

My advice is to use the examples to brainstorm the specifics of your relocation. Put your effort into breaking dependencies instead of tracking them. As mentioned earlier, don't build your plan on the premise that everything must go right. Work hard to understand which pieces can truly be done in parallel with the staffing allocated to the move. When each staff person also has another full-time job, ask, "Is parallel activity really feasible?"

Methodologies are important tools to help you drive progress for planning a data center move. I'm neutral when it comes to the methodology used *as long as you use a proven one that can deliver a successful data center move.*

Almost every methodology you will encounter is based on these elements:

- ***Discovery***—learning about the specifics of your data center move
- ***Findings***—documenting the results of ***Discovery*** into a coherent, actionable document

- ***Plan***—reviewing relocation plan, then revising and rehearsing it. (I call mine the ***move playbook,*** which was explained earlier.)

- ***Execute***—performing the plan, adapting to the surprises, and successfully accomplishing the data center move

Different outside firms will have a variety of methodologies and that's fine. They have practiced theirs and can deliver consistently as a result. However, internally led moves often have no stated methodology. Be sure to rectify that problem before any resources are expended.

Ad hoc methods lead to avoidable blunders, cost surprises, and messy moves. You're wise to avoid internal processes not specifically tuned for a data center move. They can be inefficient and ill suited to timely decisions.

I've been fortunate to work with several professional relocation firms sometimes to peer review its move plans. Here's what I've learned:

- Be open-minded and flexible about their methodologies.

- Your internal process might not be sufficient to tackle a data center move while many firms have already proven their methodologies get the job done.

- Keep a ***proven methodology*** in mind; it's the hidden driver behind constructing any successful plan.

The first WBS example that follows illustrates one way to prepare your destination. The second example shows steps to build your *move playbook*. The final example is what a discrete move event might entail and is organized by move phase. Again, these are simplistic examples; add your unique requirements.

Example #1: WBS—Prepare Destination Data Center (colocation)

This example assumes the destination is a colocation facility. If your data center is under construction, you will need significantly different components to ensure construction elements are tracked correctly.

- Assumptions
  - Destination chosen and contracts signed
  - Onboarding for access with policies completed
  - Staffing designated
  - Budget approved
  - Move governance in place
- Design
  - Logical Reference Design
  - Physical Device Design
  - Cage Layout

  - Cabinet Elevation Drawings
  - Power
  - Wire Management

- Publish Data Center Standards
  - Fiber Labeling
  - Copper Labeling
  - Circuit Labeling
  - Device Labeling
  - Racking and Blanking
  - Change Management

- Order Circuits
  - PSTN (Public Switched Telephone Network)
  - MPLS (Multiprotocol Label Switching)
  - Internet

- Build
  - Order initial equipment
  - Configure equipment to your unique requirements
  - Turn up new circuits
  - Test

- Firm Order Commitment Dates for Circuits
  - PSTN (Public Switched Telephone Network)
  - MPLS (Multiprotocol Label Switching)
  - Internet

- Deploy
  - Routing visibility
  - Pilot move
- Deliverables
  - Designs published
  - Standards published
  - Circuits ordered
- Commission Data Center
- Data Center Destination Date Certain—Ready for Equipment

Commission and test your destination data center well before moving in (as explained previously). This means thoroughly test all systems including mechanical, air conditioning, electrical, life-safety, security, and telecommunications circuits. Don't forget to test emergency power-off systems, UPS (Uninterruptible Power Supply) and generator backups, and fire and smoke detectors. Be sure to inspect for fluid leaks and other potential hazards.

CAUTION: Be sure to clean the destination data center before your move-in date. It's common to find leftover construction debris. It can ruin your brand-new air conditioner equipment as well as computer equipment. Hair-like metal fragments from debris can wreak havoc on sensitive computer circuit boards.

Example #2: WBS—Build Your Move Playbook

This example illustrates the common tasks needed to build a *move playbook.*

- Assumptions
    - Staffing in place
    - Governance in place
    - Budget in place
- Preparation
    - Gather the Master Inventory
    - Gather the Application Inventory
    - Build the Master Timeline
    - Build the Stakeholder Communications Matrix
- ***Sequence*** the Batting Order within each of the move events
- First Draft of ***Move Playbook***
    - Finalize number of moves with dates
    - Finalize Batting Order
    - Finalize Punch List for each move
    - Finalize move timeline
    - Finalize communications plan
    - Finalize move inventory
- Rehearse the ***Move Playbook***
    - Add discovered ***Go/No-Go Criteria*** to playbook

- Add contingency for discovered issues
- Brief governance team for sign-off

Example #3: WBS - Data Center Move Event

This example of a single data center move event contains all the phases. You can replicate the steps to build more complex multi-event moves.

- Assumptions
  - Staffing in place
  - Governance in place
- Cost Model construction
  - Uncertainty analysis
  - Sensitivity analysis
  - Budget iteration and review
- Site Selection
- Pre-Move Planning
  - Build and rehearse ***Move Playbook***
  - Prepare the destination
- Teardown
  - Go/No-Go Criteria consulted
  - Establish the move War Room
  - Honor the Batting Order for shutdown and teardown
- Transit
  - Follow established transportation plan

- Arrival
  - Inventory arriving components
  - Stage for re-assembly

- Re-Assembly
  - Honor the Batting Order for re-assembly and restart
  - Follow the War Room protocol for all issues

- Post-Move
  - Prosecute all issues to completion using previously established protocols
  - Celebrate staff successes
  - Update lessons learned in preparation for the next move event
  - Disposal Plan—This piece is often overlooked. Have a disposal plan for items left behind. You can unlock value by involving vendors who specialize in paying for unwanted assets. Like fruit, the longer you leave it rot in the forgotten data center, the less value you can extract.

WBS Final Thoughts

I understand the temptation to copy a WBS and then delete sections you believe you don't need. In fact, a few online websites offer examples. Given my previous three examples, it should be obvious you'll need to add specific elements relevant for your move. Yes, it takes significantly *more* work, but it delivers *superior* results.

CAUTION: When it comes to accessing my WBS library, who doesn't want to quickly copy and adapt to formulate a plan? However, I believe this approach is harmful to a successful move. Instead, I encourage you to do the heavy lifting required. You'll find the *process* is more important than the result. If you are struggling with this phase because of uncertainty, that's a sign to add additional subject matter experts to your team.

Next, let's address the critical process of rehearsing a data center move.

HOW TO REHEARSE A DATA CENTER MOVE

Move rehearsals are the best method for improving the confidence in your execution team, but this only works if you are vigilant about detecting the hand waving and false answers. Data center move rehearsals are usually met with one of two reactions: apathy or resistance. Yet these rehearsals are essential because they reveal flaws and expose areas requiring "deeper dives" to understand.

For our St. Louis client, Paul sent the *move playbook* in advance with a proper agenda and led the rehearsal with a room full of technical professionals. Knowing this wasn't his first rehearsal "rodeo," I observed quietly from the back. During his introduction, everyone stayed predictably occupied with distractions open on their laptops, phones, and tablets. Apathy was the first bull in this ring.

Then Paul asked, "Who is responsible for shutting down the e-mail server?" He waited for a timid hand to respond. After a series of questions, several participants wanted their questions about e-mail answered, too. We immediately recorded a sequence of more than a dozen steps and two offline action items.

Seeing the apathy in the room overcome, the next system owner in the hot seat resisted being questioned by others. He gave short answers with little detail—a frustrating and painful experience for everyone there. Paul persevered and guided the questioning to focus on the needed details and not the personalities.

CAUTION: The resistance shown can be as varied as the personalities involved. You can expect participants to declare rehearsals a complete waste of time. After all, they are the *professionals,* and they already know what to do. "What's the point of telling others?" they tend to ask.

TIP

WHAT MAKES A GOOD REHEARSAL?

- *Writing a clear agenda*
- *Declaring expected outcomes*
- *Seeding the audience with outside subject matter experts*
- *Encouraging dialog*
- *Recording the session for accurate follow-up*
- *Documenting action items for follow-up*

Rinse and repeat until you have created a documented teardown and reassembly sequence.

Many organizations rush or skip data center move rehearsals all together and then are surprised by the bad outcomes. Realize that complicated sequences require multiple rehearsals. Program your schedule to allow enough time to complete *several* rehearsals to remove the risk from your plan. Complicated sequences should be rehearsed at least twice.

I remember another move rehearsal that went well—too well, it turns out. The subject matter experts in attendance insisted the topic was well covered by the application vendor's documented recipes. They weren't wrong about that point, but the firm had let the support of the recipes lapse. They were no longer available to non-paying customers.

What's the lesson learned? To follow up and find all external documents referenced by your subject matter experts *before* moving day. In this case, not having support for a major storage subsystem became a huge deal that had to be corrected before going further.

Expired support contracts are not unusual. Because technology departments change, any decisions made to save money (e.g., dropping support) should be revisited for their effect on a data center move. Any missing support contract costs can also go missing in the budget forecast because the detail wasn't available during the cost model phase.

Thankfully, conducting a rehearsal can illuminate weak points in a complex plan and/or test the ability of professionals to respond to scenario-driven events. The exercises involved are necessary ways of removing surprises from complex events and gaining practice with your *move playbook*.

But how do you avoid the common mistakes when conducting your rehearsal? It's true many technical staff will not differentiate between a rehearsal and just another technical meeting in the conference room. Here's a summary of six critical lessons learned about data center move rehearsals collected over three decades. May they be lessons well remembered for your organization!

1. ***Use an outside facilitator***—The problem with using someone already connected to the move comes down to personal and technical bias. For example, a familiar facilitator may not challenge strong-minded technical experts in a productive way and may overlook key weaknesses because those experts claimed superiority on the topic. Technical bias manifests when areas aren't adequately explored due to the facilitator lacking the technical breadth to dig deeper into an issue.

2. ***Prepare the scenarios for testing in advance***—Amateur exercises end up being chaotic and possibly irritating to the senior technical staff. Rule of thumb: For every hour of an exercise, there should be about four hours of preparation.

3. ***Carefully select the participants***—Inviting everyone to get involved is like trying to bring the ocean to a boil. And inviting only technical resources can result in missing the business impacts the scenario is supposed to cover. Be sure to properly balance the backgrounds of the participants.

4. ***Divide the rehearsal into manageable sessions***—Trying to do too much leaves everyone feeling like nothing gets accomplished.

A complex scenario can be split into several sessions so lessons learned can be applied to each successive exercise. Remember, participants will likely be interrupted to manage their "real" job functions. Don't expect to capture their attention for uninterrupted and extended periods. It's not practical!

5. ***Documenting the rehearsal is more than taking notes***—Too often, the exercises conducted do not properly document the revelations in a meaningful way. Notes get distributed as if a court reporter dutifully captured every word. Instead, translate the notes into clear, actionable steps.

6. ***Do it over***—A common mistake is to declare success because the exercise or rehearsal is held once. Most internal staff members do not want to repeat an exercise and thus might even misrepresent the event as a success. But if the complexities aren't properly addressed, you face doing it over again. After all, do you practice hitting a baseball just once and expect a home run?

CAUTION: Well-run rehearsals are a lost art for most organizations, even though their value has been proven for testing complex interactions of people, process, and technology.

Does it interest you to relocate your applications to the Cloud? Let's discuss that option next.

WHAT ABOUT THE CLOUD?

Someday soon, someone will challenge you to move your entire data center to the Cloud. Enough case studies have emerged to entice executives to explore the promised economics of using the Cloud. After all, who can resist a "dangled carrot" that would reduce costs, simplify deployments, shed staff, and improve business agility?

No one. And Cloud marketing departments know that. The reason is simple: Information technology remains a cost center despite all the claims to the contrary. Turning this cost center into a value generator often ignores the complexity and legacy that define information technology applications.

CAUTION: It would be easy to discuss all of the reasons why you *shouldn't* move to the Cloud, but that's not why you're reading this. Let's focus on discussing what you should do to move your data center into the Cloud if you choose.

Prepare a Budget

Nothing happens without money, which means you'll need monetary fuel to complete your data center migration to the Cloud. Unfortunately, there are no mature economic models in which you simply enter data and arrive at a number to pitch to your CFO. If you're moving to the Cloud, you're entrusting your working applications to others. Therefore, your budget needs to reflect the labor required to engineer an application migration and, in some cases, a re-architecture that has to happen *before* a migration.

Application migrations are not new in the lifecycle of most firms. Therefore, moving these applications to the

Cloud requires the same discipline, use-case testing and certification, and skilled project management required in the past. Keep in mind the Cloud has not yet suspended the laws of how complex engineering projects must be completed successfully. Don't overlook the cost of educating and re-tooling your internal staff and users for the changes the Cloud will bring. To make sure your data will be secure, budget for the security testing required. You'll certainly incur duplicate carrying costs during your move to the Cloud, so model those carefully.

Dedicate a Team

Information technology projects fail for many reasons. Sadly, most of them are well-known lessons *learned* but not *remembered*. Expecting staff members to do their normal jobs while moving your operation to the Cloud is a mistake you can avoid. Instead, select a dedicated team to focus on the challenges, overcome them, and provide you with the best chance of success.

Define the Scope and Milestones

Like every exciting project, a move to the Cloud will ignite the demand for other projects to ride along. Some suggestions will be compelling and some will be just plain silly. Use discipline up front. That requires clearly defining the scope and expectations for completing the move and rewarding the efforts.

CAUTION: Limit scope and phase a large Cloud move to learn key lessons before going "all in."

Identify the Alternatives

If you haven't considered alternative solutions to moving to the Cloud, you are missing the opportunity to understand what you are doing. Hype about the Cloud is loud and proud. Believe all of it at your peril. Getting grounded in its alternatives allows you to effectively manage your move with a realistic expectation of what you are buying.

CAUTION: The Cloud is not magic. By studying its alternatives, you can select the Cloud components that make sense for your firm.

Determine Performance and Disaster Recovery

Understand the end user performance before you begin the migration. Do you know what your firm experiences now on every application you want to move? What additional costs are required to maintain or improve existing performance? Have you dedicated an experienced architect to review your disaster recovery in a Cloud environment?

CAUTION: Carefully read Cloud case studies, and you'll realize how much heavy lifting is involved.

WHAT'S DIFFERENT IN THE CLOUD?

Executives understand that complex Cloud migration projects require multi-disciplinary teams to increase the odds of success and realize the promised ROI (return on investment).

Unlike piecemeal approaches, a data center move to the Cloud presumes you will decommission your existing data center within a reasonable period. It also presumes you will follow through with necessary steps to realize a cost savings.

Possible costs savings include:

- ***Retiring physical infrastructure.*** *You'd be surprised how many firms continue with duplicate carrying costs of the legacy infrastructure.*
- ***Reducing both technical and administrative staff where appropriate.*** *You can't realize promised Cloud economics if staffing isn't adjusted.*
- ***Realigning technical roles and processes to provision, monitor, and service Cloud-based infrastructure.*** *The Cloud requires different skills than on-premise solutions. Realize that adapting roles* ***early*** *is a critical success factor.*
- ***Abandoning legacy applications in favor of modern, agile Cloud-ready applications.*** *Moving to the Cloud often means a legacy application gets completely replaced instead of moved. Making conscious decisions in these areas can pay huge benefits.*

- ***Allocating external resources to get evaluations done in a timely manner.*** *Contrast this with internal-only evaluations and committing regular shot-clock violations. When your internal resources are tasked to serve two masters, you can't expect a sustained sense of urgency. Dedicate staff to the project or seek additional resources to get the evaluation done so execution can begin.*

Three Differences to Expect

Let's assume you've properly evaluated Cloud vendors for security, financial stability, and desired service level agreements. What makes a wholesale move to the Cloud different than migrating to a physical data center? Quite a few things. Here, let's discuss three elements deserving your initial attention.

Performance will be different. How your applications perform might be better or worse in the Cloud, and you certainly don't want to learn this after you move. So model, measure, and test all of your applications before moving. Different applications will require different approaches to getting the performance you need. It's likely your monthly bandwidth budget will need to increase, and you might consider WAN (Wide Area Network) optimization and redundant networks. Most firms have no idea what performance is consumed by current applications because they've never been measured. Generalizing instead of measuring is certain to yield unpleasant surprises.

Incident response will be different. Your response to incidents should be well thought out and even rehearsed for every application. Your Cloud providers may not be able to give you satisfactory estimates of time to fix or even the reason for outages. Secure their incident response procedure prior to moving your applications. Some firms don't even have a formal Incident Response process documented. Such ad hoc approaches will be exposed as amateur when you move. So revise or build your incident response plan before moving to the Cloud to avoid doing it one outage at a time.

Return on Investment (ROI) will be different. Like a weight-loss program, the ROI advertised is not typical and results will vary. Plan on duplicate carrying costs while you work out the kinks. Also plan on application development projects to move those in-house, legacy applications to Cloud-friendly platforms. Set up organizational training and education to change the way your company expects to use the new infrastructure. Expect cultural changes to your support staff that could involve re-staffing the right kinds of skills. Know that resistance and hanging on to physical infrastructure will decrease your ROI the longer you cling to it after your move.

What about a rollback's effect on your ROI? Plan on a rollback if the Cloud affects your revenue or reputation. Of course, testing minimizes rollbacks, and ROI naturally elongates with more testing time.

Moving a data center to the Cloud is a multi-disciplinary project requiring project management, focusing on business objectives, and understanding the applications being moved at a deep level. Devote the necessary budget and resources to plan and execute this move for a successful outcome. Temper it with a healthy dose of realistic expectations the change will introduce.

Finally, when choosing a Cloud migration partner, find out if the vendor "eats its own dog food." That is, what percentage of its own applications occupies the Cloud versus residing on premise? Have company personnel disclose their own Cloud economic savings to counteract the hype so you can discern what is realistic.

CAUTION: Evaluate all the alternatives before selecting a Cloud service provider.

Next, let's address writing a Statement of Work.

HOW TO WRITE A STATEMENT OF WORK

Do you need to write a Statement of Work (SOW)? Engaging vendors for your move may require drafting a SOW to specify your requirements and request standardized responses—a daunting task. Has your online search for an example to jumpstart your efforts left you frustrated?

While entire books have been written on the process of writing a good SOW, you likely don't have time to read them. Worse, they don't include the kind of details you'd like to have for your firm's data center move.

Let's look at the two key questions involved.

1. What are the minimum SOW elements?

Complete these basic tasks for every Statement of Work:

- Define the Scope
- Define the Deliverables
- Define the Timeline and Period of Performance
- Define the Evaluation Criteria
- Define the Reporting Requirements
- Add your Terms and Conditions

2. What data center relocation elements should you include?

These questions can help you get specific about your project:

- What are the origin and destination locations?
- What total square footage is in each location?
- How much equipment is moving? (Provide the inventory list to save time in vendor questions.)
- What downtime is acceptable?
- How would the work be divided between your staff and the vendor?
- Do you have a master move timeline?
- Do you have special equipment to move?
- Do you need specific technical expertise?
- Do you have enough time to run a competitive process, or do you have to get a vendor on board quickly?

To frame your knowledge of the issues to be solved, include your answers to these questions in your SOW. Don't be afraid to outline areas of uncertainty so vendors can demonstrate their experience solving those issues for you.

CAUTION: Without knowing your specifics, it's impossible to list all the critical details that would make your move successful. To help evaluate your list, have your SOW reviewed by peers before releasing it to your vendors.

Is the Perfect SOW the Answer?

Because a data center move can be complicated, having a well-written SOW represents only one step in this complex journey. Ultimately, your success depends on multiple elements, not the least of which is allowing enough time to plan your move.

It takes time to review all the vendor responses and ask clarifying questions. Guard against using a disproportionate amount of time and resources at the beginning of the process and leaving little to the practitioners—the people who will be tasked to execute the plan and actually move the data center.

You might be tempted to dump the task on people in your procurement department. If you do, what's likely to happen almost immediately? They look online for boilerplate templates and issue a SOW devoid of your specific needs. How do I know this? I have a library full of badly written SOWs for data center relocations collected from decades of moving data centers. Responding to these is painful for all involved. Vendors don't know what you need when you can't articulate those needs well.

CAUTION: Using a SOW to fish for ideas on how to move a data center is incredibly inefficient. Do you want a Frankenstein monster of a plan built from different vendor responses?

WHAT CAN GO WRONG

Let me illustrate the dangers of one particular SOW process. The firm's procurement department wrote a generic SOW and collected vendor responses. Vendors also submitted written questions that were answered and returned to all participating firms. The questions themselves exceeded the length of the original document! And the answers given to the vendors were inconsistent! No wonder the firm missed its deadline for vendor selection.

Ultimately, the SOW was withdrawn.

What happened to the data center move in the meantime? The firm simply contracted one of its existing vendors with a hasty time and materials agreement. By waiting until they had run down the shot clock to almost zero, they placed urgency on the vendor to perform in a crunch—an avoidable, expensive blunder.

To have an understanding of all phases of a data center move, keep reading this guide. With this information, you can tailor your SOW to the areas requiring assistance and avoid creating a generic document and sub-par vendor responses. Remember to peer review your SOW before you release it to vendors. That extra effort will help you get the most from the process.

DAILY PUNCH LIST

In 1999, I had the privilege to consult during the construction of the new San Francisco International Airport (SFO) terminal project. This massive undertaking at one of the world's busiest airports came with every imaginable complexity. Keeping track of all the construction details as well as the sequence of events and their dependencies couldn't be done with a single tool. An entire army of project-management specialists, a wide range of tools, and endless status update meetings with diverse groups ruled this complicated project.

For example, subcontractors began their morning with a mandatory coordination meeting. Held in a hot, dusty construction trailer, the general contractor outlined the sequence of events for the day's work to standing room only. A choreographed plan emerged daily of how each subcontractor would participate—or stay out of the way—to permit the task's completion. Essentially, conversations turned into agreements of cooperation or disagreements for escalation.

After a few months, I learned that some subcontractors consistently performed their tasks on time and others struggled with their work queue. After shadowing several of these groups during the project, I was surprised to find what set these contractors apart. While they all used the automated project-management tools required by the general contractor, the most effective subcontractors used something extremely simple to drive their productivity—a daily punch list.

Daily Punch List Provides Focus

The daily punch list is a single sheet of tasks for completion. Assigned at the beginning of the shift and returned at the end, complicated Gantt charts didn't find their way into the field. Supervisors hyper-focused on the tasks and a separate team fed the giant project-management machine of the SFO airport master plan. The electricians completed their work beside welders, carpenters, and painters. No one stopped an electrician during her workday to pull her into a meeting and grill her on percentage complete or have her predict when she would finish tasks.

When it came time to install the information technology for SFO, the prerequisites had already been completed efficiently through the use of daily punch lists. I emulated the punch list process for my special systems team working to install and test the necessary technology that operates airports.

CAUTION: For information technology projects, do you cling to serving the project-management tool du jour? Do you feed it constant status and want a perfect prediction of the future from it? Consider this: A specialist's time is wasted in status meetings and variance reports and re-planning and guessing and being punished for inaccurate guesses. A wasteful practice gets repeated when we strive to *serve the tool* and neglect the actual work.

THE POWER OF THE DAILY PUNCH LIST

As you struggle with your data center relocation project, tap into the power of using a daily punch list. Instead of focusing on manipulating the latest technical tracking tool, focus on the work at hand.

I recommend beginning with a punch list before you move to a project-planning tool such as Microsoft Project. What makes a good punch list? After studying the punch lists of all the subcontractors at the SFO project, I developed these seven categories that every punch list should contain:

1. ***Status**—This numeric scale of 0 to 4 indicates the disposition of a particular item. 0=Not started, 1=Serious issues of schedule or resources or constraints, 2=Some concerns, 3=Task progressing, 4=Task completed*
2. ***Item #**—A numeric integer assigned to each specific task*
3. ***Description**—The brief task description*
4. ***Next Action**—What's expected to happen next*
5. ***Owner**—The primary owner of the task*
6. ***Possession Arrow**—Declares when the owner needs another resource to move the task forward; drastically improves the dynamic of getting tasks completed*

7. ***Due Date****—For tasks in which certain milestones affect the overall project. Note: Assigning due dates for every task is counterproductive. Focus on completing work instead of tracking changing dates. If you use due dates for the most important tasks, not the minutia, you'll already be light years ahead of most project managers.*

What goes on your initial punch list?

- Start by listing all the disconnected tasks swimming around in your mind.

- Reduce your anxiety by emptying everything concerning your data center move onto your punch list.

- Note prescriptive tips in every section of this guide. When a tip matches your situation, put it on your punch list, then assign an owner and next action. Once you get through this guide, you will have an actionable punch list.

Punch List Dashboard

(Summarizes individual items)

QTY	Status Key	Description				
4	0	Task Not Started				
1	1	Red - Serious (schedule, resources, constraints)				
0	2	Yellow - Some concerns				
0	3	Green - Task is progressing				
0	4	~~Task is complete~~				
Status	**Item #**	**Description**	**Next Action**	**Owner**	**Possession Arrow**	**Due Date**
2	1	Establish Governance	Identify decision-makers	You	You	
1	2	Build a Cost Model	Identify known and unknown cost elements	You	Budget Pro	
0	3	Gather the Master Inventory				
0	4	Gather the Application Inventory				
0	5	Build the Master Timeline				
0	6	Build the Stakeholder Communication Matrix				

Figure 2- Example Punch List

Because the punch list is such an important tool, I'm providing a link to retrieve my punch list template. Grab it now and fill it in as you read this guide or bookmark it to retrieve it later at the moment of your need: http://datacentermoving.com/punchlist.html.

CAUTION: Adults differ in how they learn. In my workshops on data center moving, I discovered busy professionals are mostly distracted even when they're physically present in the classroom. Long after students completed the workshop, I'd receive e-mails to help them find topics covered in class. That's the reality today. You can't possibly absorb everything and recall it perfectly at the moment you need it.

PRE-MOVE CONCEPTS IN REVIEW

In this Pre-Move Planning section, let's review the major sections already addressed before moving on to teardown phase. Major concepts include:

- *Flawed planning methodologies*
- *Importance of breaking dependencies instead of simply tracking them*
- *Components of a* ***move playbook***
- *How to build your* ***move playbook***
- *Time boxing technique for your Master Timeline*
- *Examples of practical Work Breakdown Structure (WBS)*
- *How to rehearse a data center move*
- *Basics of moving to the Cloud*
- *How to write a statement of work (SOW)*
- *Importance of a Punch List*

It may seem as if you'll never complete pre-move planning, but no doubt the teardown phase will arrive before you're ready. Let's discuss it next.

TEARDOWN

DON'T RUSH TEARDOWN

The day has come to shut down and tear down your equipment in preparation for your data center move. How are you feeling?

Just as I did on the plane to St. Louis, that's when I tend to review the *move playbook* and my rehearsal notes up to the last minute. Specifically, I'm searching for missing items.

During one of my first moves, I recall how well the preparation had gone. We had progressed through about half the equipment when the phone rang. The technology director called to say, "We need to make up some time. The boss wants everything back up sooner than planned."

It's easy to know how this ended. We abandoned the discipline of following the rehearsed move plan and took shortcuts instead. Abandoned rehearsed checklists laid coiled like unused safety ropes beneath a climber on a cliff. The effort to clean up the mess created by the rush far exceeded any time saved advancing the teardown phase—a critical lesson learned. Will ours be a lesson *you* will heed?

Follow These Simple Rules

This teardown phase of your data center move should follow a few simple rules.

Make sure everything is labeled well *before* you begin teardown. Attempting to label and document *during* teardown produces a rushed and sloppy label or no labels at all.

Freeze all changes to the data center systems during teardown. This is not the time to add a new storage array, change out a fan, or install the latest kernel software. Trying to kill two birds with one outage stone introduces significant risk to the move. You may miss both birds.

Don't rush teardown. Be sure to identify in advance all specialized tools needed for physical disassembly. Your staff needs the proper power and hand tools to complete teardown safely and without damaging the equipment. People break things when they are in a hurry, and one damaged proprietary cable can ruin your entire move. With your Batting Order already documented, your task in the teardown phase is to enforce its compliance. **Note:** A teardown can happen in more than one place simultaneously (e.g., remote staff might do the shutdown of the equipment before the physical teardown).

Ask key questions. During teardown, how will you maintain collaboration and communication between remote and onsite staff? How will you prevent teardown from affecting other components of an active data center? How will you recover from the inevitable mistakes and unforeseen issues?

Those are only a few of the questions to consider. Your rehearsals will inform you of other items specific to your move.

PREPARATION STEPS FOR TEARDOWN

At a minimum, consider these preparation steps for teardown:

- Designate a Move Commander.
- Publish your Rules of Engagement.
- Publish a Master Contact List including vendors.
- Arrange for a dial-in phone bridge.
- Have a physical War Room.
- Use online collaboration tools for screen sharing.
- Plan for deploying rescue swimmers.
- Set up real-time access to system and network monitoring to alert for unintended consequences from shutdowns.
- Test communications in the data center and use radios if cell reception isn't adequate.

Communicate your teardown status to your stakeholders at regular, pre-established intervals such as every three hours.

Several steps remain requiring vigilance so make frequent updates to keep your team engaged.

Rescue Swimmers

I'm often challenged when I insist on budgeting for rescue swimmers—the resources you need to have on standby. Like the United States Coast Guard rescue swimmers, they are highly trained, skilled, and *fearless*—exactly the type of resources you want when your data center relocation is sinking fast and your staff is exhausted.

The skills of your rescue swimmers will differ based on the depth of your own technical bench. In fact, there's no reason your rescue swimmers can't be internal resources. Arrange for these swimmers in advance.

CAUTION: Most data center moves happen after business hours, and your ability to find qualified experts at that time is limited, so plan for that.

Screen-sharing Tools

Another shortcut is failing to use readily available screen-sharing tools. Shutdown of applications is a technical step. Having many professional eyes on the sequence will aid in starting up systems later. Don't worry about your peers looking over your shoulder; let the entire team help make your move successful.

Finally, heed the warning to designate a *Move Commander* to orchestrate the teardown. A command and control structure is the most effective way to keep a data center move on track.

I recall teardowns that went well and others that burned like a pile of twisted metal on a runway. In hindsight, the root causes seem preventable, but our society seems to value fire fighting over fire prevention.

TIP

TEARDOWNS GOING WRONG

Here are several root causes to avoid in the teardown process:

- *Staff deviates from the rehearsed script and rules of engagement*
- *Not enough labor to meet deadlines resulting in mistakes*
- *No process to prosecute complicated issues*
- *Miscommunication triggers cascading blunders*
- *Contingencies not deployed to save costs*
- *The move's complexity underestimated*
- *Shortcuts unwisely taken to appease last-minute demands*
- *Saboteurs who may be internal to your firm*

What is the number one factor in a successful teardown? *Rinse and repeat your move rehearsal to fully document the Batting Order and shutdown sequences.* Rehearsal is the practice of fire prevention—one that avoids the need for firefighting.

Virtualization

Virtualization provides additional capabilities meriting exploration. The simple explanation of virtualization relates to physical machines being turned into virtual workloads as files representing those physical machines. This permits running these virtual workloads on fewer physical devices. And these virtual workload files can be transported over the network or onto a removable disk and then physically shipped. Moving entire machines over the network is possible and preferred in some cases. However, not every firm has the required bandwidth between the origin and destination to make this choice practical. You will likely mix physical moves with data migration to include moving virtual machines over the network. Your actual timelines will be determined by the empirical testing of transfer rates.

Data center moves can be the catalyst for considering virtualization. Explore this option seriously; the economic benefits are substantial. Consolidating your physical workloads to virtual workloads first and then moving the virtual workloads as a data migration dramatically reduces the number of physical pieces to move. It does not, however, reduce the complexity of your move. It simply changes where those complexities occur.

Keep in mind you are moving applications critical to your firm's business. As such, testing the virtual workloads in a production environment is required before you take the step of moving these workloads. Remove most of the risk of the move *prior* to the move itself. That way, you gain certainty about how your applications will perform and what tuning will be required *before* you migrate them to the desired destination.

Virtualization introduces the option of continuing to run on the physical equipment while preparing the virtual environment. In those circumstances, account for the data change happening *after* the physical machine has been converted and *before* the virtual machine is activated.

Different platforms require different strategies for data change. Relational databases have built-in capabilities for maintaining replicated copies or for restoring changes that happened from a given point in time. File storage synchronization can be done multiple ways including specialized software to maintain exact duplicate images. Be sure to inspect each application to answer the data change question *before* moving. For example, Microsoft Exchange, a common on-premise e-mail solution, has a few built-in methods for migrating mailboxes.

CAUTION: Most firms have ridiculously large mailboxes, making migration take longer than predicted. The solution? Attempt to have users trim their mailbox sizes prior to attempting a mailbox migration.

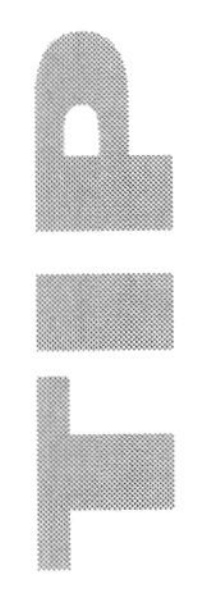

STEPS OF A VIRTUAL MIGRATION

At a high level, a virtual migration requires these steps:

- *Convert physical machines to virtual machines*
- *Transfer the virtual machines to the destination either over the network or by shipping removable drives*
- *Synchronize the data change that occurs up to the cutover time*
- *Power down the physical machine and complete the final data synchronization to the virtual machine*
- *Power up the virtual machine*
- *Validate and test the result*

The virtualization technology you use will dictate the features available for migration. While most applications are suitable for virtualization, due diligence is required to avoid any surprises that virtualization might introduce. For example, a commonly overlooked area is the proper configuration of external storage necessary to host the virtual machines.

CAUTION: Ensure you have the expertise to have this optimized for the virtual workloads you are hosting.

Hybrid Moves

Many data center moves have become hybrid physical moves combined with virtual moves. The technology of virtualization is feature rich and beyond the purpose of this guide. While virtualization can be complicated, many vendors can provide the catalyst you need *before* your data center move.

Some firms attempt to use the free licensing options and miss out on many features that substantially aid a data center move. My advice? Don't step over dollars picking up pennies when evaluating paid licensing features.

Next, what do you do on transit day?

TRANSIT

WHAT TO DO ON MOVING DAY

Typically, I don't sleep well in a car. But I was tired and Paul had the wheel so I closed my eyes and rested. I'm unsure why I woke up; maybe I heard a thump. The moving truck with all the equipment was several car lengths ahead, and the sky still dark with only a few stars punctuating the clouds. Traveling on a narrow, two-lane blacktop road in a rural area of Missouri still a few hours from dawn, Paul later recounted the drama I had slept through.

The truck ahead had suddenly braked and swerved wildly several times. The vision of pulling computer equipment out of a smoking, twisted wreck in the ditch caused him to jump on the cell phone, curse the driver to slow down, stop swerving, and forget about braking for the rabbits.

Rabbits? In Colorado, we watch for deer crossing the road all the time or those pesky falling rocks in the mountains. But certainly rabbits don't run in herds, right? Paul insisted there had been many, many rabbits crossing that dark, narrow road in the past hour.

"Did you kill the Easter bunny?" I'm unsure I really wanted to know.

Refusing to answer, Paul stopped the car and we switched places. I imagined ears and tails everywhere as my eyes darted wildly back and forth across the road. It was Paul's turn to sleep peacefully. I swear a wry smile appeared where a scowl had been hours before.

It wasn't the first time we followed the equipment truck to the destination, and it wouldn't be the last. When Paul and I rehearse the transit plans with our clients, we don't consider them complete until we've contemplated the contingency for rabbits.

Packing the Truck

If you undertake the transit yourself, understand how to properly prepare, pack, and move data center equipment, paying particular attention to warranty details. You should also know your insurance options before you reach this phase. If you use a vendor, review its procedures for keeping your equipment protected. Ask questions and secure commitment to use documented procedures.

Always ask if your equipment is subject to additional loading and unloading. This can happen for a number of reasons. Obviously, I recommend you stipulate it *never* happens, so be sure to include that requirement in your contract. A professional moving company that moves electronic equipment will have a documented set of policies and procedures for its employees to adhere to. Ask to see these and inquire about the training its staff undergoes.

Typically, I visit professional movers on behalf of my clients. Most are highly organized and take pride in the level of training for its staff. By engaging a mover early, you can benefit from its previous experience with large moves. Ask if special requirements such as chain of custody, tip and crush sensors, and team drivers can be accommodated.

Take additional care with tape storage media because changes in temperature and humidity can result in failed media. In particular, packing tapes improperly is a common mistake that can leave you with damaged media and data loss.

CAUTION: Conduct several origin inspections throughout the transit phase to catch important items left behind.

WHAT A TRANSIT PLAN INCLUDES

Whether you're moving the equipment yourself or using a professional moving company, your documented transit plan should include the following:

- Adequate time to properly pack and load the trucks. (Don't skimp on labor; more personnel means better care of the equipment.)
- Previously approved optimal route to the destination data center and recommended alternative routes
- Contact information for everyone involved in the actual move, including cell phones and escalation paths
- A plan for bad weather, traffic jams, and other contingencies
- A Move Manifest that lists everything going on the truck. The carrier often lists equipment but doesn't include the detail you need. Use this list to confirm everything gets delivered at the destination; it's your responsibility.

Do-it-yourself Packing

Why people insist on do-it-yourself packing baffles me. On your first move, why would you trust your equipment to packing methods you found online? Packaging servers for shipment through common carriers carries its own risk and requires special care.

The bare minimum procedures include these requirements:

- Package all servers individually.
- Remove all USB dongles, cables, and other attachments and ship these separately.
- Use anti-static bubble wrap.
- Know the proper box sizing and corner protection methods.
- Understand when to use advanced techniques.

I strongly recommend you leave packaging and transit to professionals. They could fill pages of this guide with their expertise. Plus, by using experts for this phase, you can conserve your resources for the important re-assembly steps.

Next, what can you do to make the arrival go smoothly?

ARRIVAL

STAGING COMPONENTS DURING ARRIVAL IS KEY

Equipment may arrive at the destination data center out of sequence and possibly in stages. Thus, you should designate a staging area for the equipment and inspect it upon arrival. Group the components in a staging area and release them for re-assembly when you've determined all the components are on hand.

As you did during teardown, be careful not to rush to the re-assembly stage prematurely. Trying to make up for time lost in previous phases can result in dropped equipment, misplaced cables, and avoidable chaos. Not acceptable.

Have on hand enough or more labor than you think you need and don't overcommit to unrealistic last-minute promises made in previous phases. Keep track of the equipment arriving and the equipment still in transit. Were these types of real-time adjustments discussed during your rehearsals? They should have been.

CAUTION: It bears repeating. If your plan hinges on everything going right up to this point, you may be blindsided by hazards.

When Paul and I finally arrived early Sunday morning after following the equipment truck dodging rabbits, we were relieved when we saw it backing to the receiving dock and the trailer doors opened. A quick inspection inside the truck showed no shifting or damage had occurred. So far, so good.

"Who has the keys?" Paul called out, wanting to get inside the building to open the dock door.

Silence told us they had forgotten the keys, even though they marked that checklist item complete during teardown. In fact, the keys were 10 hours away by car back at the origin. Paul did a quick perimeter check around the building with a tire iron in his hand looking for another way in. It was less expensive to break a small side window than have all the resources wait and delay the move another 10 hours.

No plan survives contact with reality, and your move will be no different. Something as straightforward as arrival at the destination might present a challenge—even if it was rehearsed!

Elements to Check at Arrival

Common elements to check in the arrival phase include:

- Building entry. Remember many sites have strict access control in place and facility entry permission can take weeks. Classified facilities take even longer.

- Floor and wall protection worked out in advance

- Allowed dock and elevator usage. Many multi-tenant buildings restrict usage after hours.

- Landlord contact numbers to call if conflicts need resolving

- Weather issues in case the equipment might be exposed during unloading

- Disposal of packaging materials. Know where accepted locations are; don't assume nearby dumpsters can be used for your debris.

What's next? Re-assembling the equipment.

RE-ASSEMBLY

WATCH OUT FOR STAFF FATIGUE

Using the same people who performed the teardown for re-assembly increases your chances of success. One critical error, however, is failing to plan for their rest breaks between those two events. You don't want your people to be tired at the precise moment you need their sharpest focus. It's best to use the transit time to rest them.

Keep a critical eye on your high performers; they are likely to push themselves throughout the move process. Remind them the importance of pace. Like professional basketball stars Michael Jordan and LeBron James, you'll want them alert on the floor during the fourth quarter of your move.

During re-assembly, walk the data center and observe the re-assembly progress. Ensure staff members follow good practices for resolving problems. A critical lapse of judgment—such as allowing "dart-throwing" to substitute for systematic troubleshooting—can turn a simple problem like a bent pin on a cable into a spiral of compounded errors.

Plan for Acclimating Equipment

Acclimation is letting equipment adjust to changes in temperature and humidity. Many equipment manufacturers do not properly address acclimation—either giving no guidance on durations or giving ridiculously long acclimation times. So use common sense. If you have special equipment, require the vendors to document in writing their recommended acclimation times so your timeline can be adjusted. To minimize the acclimation time, use climate-controlled trucks during transit. Don't simply skip this requirement!

Use extra care where you have older, spinning disks that have not been powered off for a long time. Equipment mortalities can be high for these devices and it might not show up until a few weeks after the move. Use the acclimation time to muster your staff, review the reassembly procedures, and make adjustments to outgoing communications.

Know your own limits and those of your staff. Be prepared to escalate problems beyond your staff to meet your re-assembly timeline. Have a stand-by team of rescue swimmers to save important time during a crisis.

PUTTING HUMPTY DUMPTY BACK TOGETHER

Re-assembly is my favorite phase of the data center move process. There's an excitement to orchestrating the successful re-assembly of all the components—finally piecing Humpty Dumpty together again.

My approach is to establish specific Rules of Engagement well beforehand. These basic rules reinforce teamwork and discipline during the re-assembly phase.

1. Practice disclosure during all phases.
2. Avoid unilateral decisions. Stick to the script.
3. Use the War Room process for all issues.
4. Maintain your perspective!

1. Full Disclosure

This means reporting all anomalies during any phase, such as an accidental power-off prior to a graceful shutdown. This avoids unnecessary or incorrect steps during re-assembly springing from inaccurate information. It allows time to consider alternative steps required to successfully restart that component.

Understand that full disclosure isn't a natural habit for technical staff who fear blame will be assigned to them. Overcome this fear of your team since full disclosure is critical to prosecuting the re-assembly with the right information.

2. Unilateral Decisions

Making isolated decisions is another difficult habit for technical staff to break. Solving issues on the fly in an ad hoc manner without consultation should not go unchecked. Ensure all issues reach your War Room to avoid unintended consequences affecting other efforts. A simple item such as bringing a database online out of sequence can result in data loss due to out-of-sync information, thus preventing the roll-back process from working.

3. War Room

The War Room becomes your nerve center and mission control during the re-assembly phase. Designate a Move Commander to run the War Room with strong, decisive leadership. Think back to your rehearsals. Who exhibited the traits you need for a Move Commander?

It's unclear who can take credit for the term War Room. We know Winston Churchill used war rooms as command and control centers to make decisions in real time during the Second World War. The U.S. space program uses Mission Control to orchestrate complex sequences. In the context of a data center move, the War Room can be physical or virtual.

The critical elements of running a successful War Room are:

- All deviations from plan are reported.
- Resource reassignments are made in response to inputs.

- Escalations are communicated.
- A ticketing system is used to maintain order.
- Regular status gets reported to stakeholders.
- A Move Commander is in charge, not a committee!

4. Maintain Perspective

Many data center moves extend well past intended timelines. Staff members lose patience and get fatigued. The personalities are diverse and so are the responses to exhaustion with professionalism taking a back seat to brutal honesty. Perspective can moderate this decline. Remind participants at regular intervals to maintain their perspective and intervene quickly when tempers flare.

Do you remember my advice to have your rescue swimmers on standby? Don't be afraid to use them! Get problems solved quickly. Don't let experimentation deceive you into thinking trial-and-error methods are less expensive or quicker.

CAUTION: You're on the clock to get your data center operational. Solve for that problem and don't focus on avoiding costs during this phase.

Not Down Yet!

In Colorado, while driving down the Rocky Mountains on Interstate 70, you'll see signs warning you of the steep grades. However, the highway plateaus when the skyline of Denver appears and you relax believing the dangers are in the rear-view mirror. Just then, a large, yellow overhead sign appears:

TRUCKERS YOU ARE NOT DOWN YET,
ANOTHER 1 1/2 MILES OF STEEP GRADE
AND SHARP CURVES TO GO

The runaway truck ramps appear as the road increases in grade and winds through more curves. If truckers do not heed that sign, their speed will be dangerously high at the precise time the hazards are encountered. As expected, some do ignore this warning and suffer as a result.

Data center moves should also come with a sign warning:

YOU ARE NOT DOWN YET,
MORE CHALLENGES AHEAD

Even when re-assembly is finished and you've sent everyone home, you are not finished. Unlike the mountainous interstate highway, there will be no runaway truck ramps to catch your mistakes. Post-move surprises color your firm's perception of a move's success or failure so be vigilant in the hours and days following a data center move.

Let's look at some of those post-move issues requiring your attention.

POST-MOVE

TECHNICAL AND ORGANIZATIONAL ISSUES AFTER A DATA CENTER MOVE

Is it surprising that data center moves uncover organizational issues in addition to the technical challenges encountered? Yet, many are unprepared for this painful exposure.

Post-move tasks are rarely forecast correctly in the budget, which results in ad hoc responses. From a personnel standpoint, staff losses during the post-move period are common because of staff burnout from working two full-time jobs simultaneously and a lack of recognition.

Giving retention bonuses, while popular, may backfire because they serve as a perverse incentive to stick around rather than to achieve excellence during the move. Feelings of resentment toward management can be acute during the post-move period.

It's important to avoid these two critical staff recognition blunders:

1. Failing to recognize individual contributions with specifics that made the data center move successful.

2. Giving blanket recognition in which everyone is congratulated for a move well done. This offends the true heroes of the move and proves management wasn't paying attention. You risk re-igniting old prejudices staff may harbor due to similar management behavior in the past.

This kind of management laziness is often the catalyst for the under-appreciated to seek new opportunities. But it is so unnecessary! What can you do?

- Involve your human resources department in advance so recognition criteria and bonuses can be well planned, funded, and executed.

- Tap into the motivational value of food, clever T-shirts, and a genuine, unexpected "thank you" to keep morale high.

- Find a variety of ways to show the staff you're paying attention.

CAUTION: Managers can also be tempted to leave following a data center move. Executives need to monitor post-move organizational disruptions closely.

SURPRISE THEM!

During one move, the destination data center preparation was compressed into a few days because equipment arrived late. By the second day, a small staff of engineers doing the heavy lifting of racking and stacking new equipment was behind schedule and physically spent. I led one member out of the data center into a room with a waiting chair masseuse. In turn, he selected the next member to receive a surprise chair massage until the entire team had received treatment.

That team still remembers those massages that showed up ***exactly*** *at their moment of* *need. Contrast that against a few impersonal gift cards arriving months later to everyone. Pay attention and act to deliver timely, sincere, and **targeted** gratitude.*

Consider these additional post-move checklist items:

- Is there a budget for post-move tasks including recognition events and key contributor bonuses and raises?

- Are managers tasked with identifying the true key contributors throughout the long process with specific examples so their performance does not go unrewarded?

- Is your staff properly augmented with skilled resources to avoid burnout and resentment?

- Is proper attention being given to post-move custodial items such as decommissioning circuits, equipment, and facilities?

CAUTION: What about your customers? Are you listening for unintended consequences from the move and actively confirming there are no issues? Customers today are quick to take to social media to air their concerns. Give them a proactive, productive outlet instead. Task your firm's communication department with this responsibility.

Neglecting data center relocation post-move tasks can taint an otherwise successful data center move. With the proper attention, you can avoid these hazards.

Decommissioning

Often neglected, forgotten, or abandoned is the post-move step of decommissioning. It's a big effort incorrectly assigned to technical staff who have to keep the moved data center up and running.

What is decommissioning? It's the proper retirement and disposal of the data center and all of its remaining systems. Factors include power and cooling, fire suppression, generators and batteries, and left-behind hardware assets. There are also financial, regulatory, and environmental factors in play with a data center decommission.

At a minimum, these areas should be addressed when decommissioning your data center:

- Data destruction with chain of custody verification
- Information technology asset recovery or disposal (with accounting in tow for financial compliance)
- Facility-based asset recovery or disposal to include items like computer room air conditioners, fire suppression, generators, and batteries

- Abandon, dispose, or sell cabling plant, raised floor, and/or rack furniture

Many risks exist in this post-move step of decommissioning. While staff should be able to carry out the firm's data destruction policies, hire a data destruction firm to do certified data destruction more efficiently with less cost than assigning these tasks to internal staff.

Decommissioning is a necessary post-move step that's best executed by an experienced team reporting to the chief information officer, chief financial officer, and your facilities team. Choose a vendor partner that understands the secondary markets for all of the components in your data center. Too often, big-ticket items such as generators and battery systems are ignored while too much effort is spent trying to re-market low-value technology assets.

Typically, the technical staff is not suited for what decommissioning requires. Do they work with the environmental recycling requirements or any applicable legislated compliance requirements? Do they understand the proper accounting required to retire assets carried on the firm's books?

CAUTION: Think of decommissioning like quicksand. It can slowly swallow those who flounder to decommission the data center, so keep your technical people focused on running the newly moved infrastructure smoothly. Use other resources for decommissioning.

Paul and I were exhausted. After a long drive, the St. Louis airport exit drew near. Another new data center move, already waiting, required immediate attention for an urgent timeline. I tried to push it out of my mind. I kept thinking if Paul gets the middle seat again on the flight back to Denver, there will be trouble! From his sour expression, I knew he already thought the same thing.

I checked the time. A small margin existed for returning the rental car and getting through security right before the flight attendants closed the cabin door behind us. Everything had to go right to make that margin.

But then our plans went deliciously sideways.

A billboard's bright message caught us both by surprise. A sizzling bone-in rib eye on a pristine white plate beckoned us. We took the next exit and found the restaurant easily. Inside, we felt underdressed and slightly embarrassed tucked away at a small table in the back without reservations. However, the hot, white plates arrived and we savored the aroma before tasting. Brushed with melted butter and perfectly trimmed, our full-flavored steaks did not disappoint.

While we missed the flight home, we experienced a memorable and satisfying celebration of our St. Louis data center move just completed. With the accomplishments still fresh in our minds, that simple steak dinner did more than boost our morale. We shared lessons learned. We laughed remembering details that seemed so serious a few days before. We finally made it all the way down the mountain after months of preparation—navigating successfully through surprises and hazards. Neither of us has forgotten that accomplishment. It's linked forever in our minds with the term "perfect ending."

TAKE TIME TO CELEBRATE

I want you to feel that satisfaction of making it safely down the mountain. Celebrate your success with your team in the days, not weeks or months, following the completion of your data center move. People, not computers, are the most important piece of the data center move puzzle.

Don't be lazy by simply sending the generic "all-hands" e-mail thank you. Instead, plan an in-person celebration worthy of everyone's accomplishments. Appreciate what they did by listening to them and celebrating heartily. Harnessing the morale and extraordinary effort of the staff whose teamwork made you successful is good business. Trust that your investment in your celebration will be repaid frequently in unexpected ways.

The following *Executive Quick Guide* will help you bring on board your busy executives. Right to the point, it lists actionable steps and key takeaways for those who seek to understand the hazards ahead. Every data center move benefits from having an informed and involved executive team. The next section helps achieve exactly that.

EXECUTIVE QUICK GUIDE

PROVEN TIPS FOR BUSY EXECUTIVES

Coaching executives during their data center moves has given me insight into their challenges. As an executive reading this guide, here is my straightforward advice to you. It represents truth to power based on two decades of coaching executives.

I suggest:

- Before you hire external resources, make some decisions. Costs are best controlled when you don't boil the ocean of possibilities.

- Instill a common sense of urgency across your entire organization, not only inside the technology department.

- Help your staff members escape the organizational coefficient of drag. Keep the project managers out of the wagon that your staff pulls up the hill.

- Challenge all plans that call for everything having to go right.

- Guard against preventable staff losses in the post-move phase.

- Use a successful move as momentum to fuel further changes.

- Start early and don't waste your shot clock by underestimating the level of effort to reach escape velocity. You know your firm the best. How long does it take to go from decision to execution? Starting early helps people do their best at critical times.

After one successful data center move, I was invited back a few months later to fill the interim role of director of information technology while the firm performed a search for a permanent replacement. Here's what I found:

- A new set of challenges in a new facility had stretched the team thin.

- Several staff members had accepted new opportunities and moved on.

- Instead of one unifying issue—moving a data center—there were dozens of urgent technology projects to finish with scarce resources during this time of disruptive organizational change. This tells me that the post-move phase remains the most challenging part of a data center move.

I can promise you will work extremely hard to get your team ready for your data center move. It's much easier to *maintain* that momentum after the data center move than to *restart* it.

Realize that a data center move touches your entire business, thus creating a unique opportunity to influence people differently and change their perceptions of your team. I encourage you to think beyond the data center move. What projects will consume your team immediately upon completion?

While your staff executes the data center move, focus on the new resources you will need to avoid a devastating staff burnout after completion. Keep track of what went well and what went poorly within your organization during the data center move. Because the move exposes existing dysfunctions, you are in a great position to apply corrective action for a chronic condition that has escaped scrutiny. Recognize that not all dysfunction requires a technological solution. Changing behaviors starts with people.

A move can expose critical dysfunctions that had existed long before the project began. Watch for:

- A lack of professionalism in which business units undermined other units to advance their goals.

- Imprecise communications that resulted in wasted time to come to agreements.

- Unclear ownership of tasks slowing progress and complicating oversight.

- Unwillingness to take individual responsibility as a way to avoid penalties.

None of these dysfunctions require a technical solution. Rather, they require the hard work of executives to confront poor behavior and devise acceptable methods that change those behaviors.

A data center move will certainly expose them, so be ready, be courageous, and be a leader.

CAUTION: Everyone brings their pre-concieved ideas to a data center move project. Reset the focus to your objectives with a systematic approach to onboarding using this guide as a start.

KEY TAKEAWAYS

REVIEW THEM OFTEN.

Remember, lessons learned are not lessons remembered *unless* they're rehearsed. So review these key takeaways:

- Create a ***move narrative*** early in the planning process.
- Anticipate onboarding at inopportune times and mitigate that using this guide, your own lexicon for your move, and ongoing iteration of your ***move narrative***.
- Develop a cost model, not simply a budget number.
- Calibrate your narratives and cost models with an uncertainty gauge. This will aid in peer reviews and focus your scarce resources on high-risk items.
- Remember that narratives inform the interpretation of numbers in a spreadsheet.
- Make some early decisions and establish an overall sense of urgency.
- Assess your needs for outside help honestly and realistically.
- Communicate coherently to your vendors with consistency.

- Over-communicate during all phases. Customers, vendors, and staff all need care and feeding during a data center move.

- Anticipate hazards by rehearsing your move, your problem resolution protocols, and your communication strategy and messages.

- Don't rush the teardown and re-assembly phases.

- Plan for acclimating the equipment.

- Avoid recognition blunders. Reward your staff!

With so many technical details embedded in each data center move, it's impractical to cover them all in this guide. Avoid the "tyranny of the technical" from distracting you from using sound principles to manage your move. Remember, people are the most important component of your move—and they are infinitely more complex than machines.

To illustrate this point, let me leave you with six common factors than can undermine data center moves.

1. Not creating a sense of urgency.

Create a sense of urgency. Without it, there's no urgency at all! A sense of urgency is essential in any organization preparing for a major move. Often, you'll find there is no shared sense of urgency—or worse, no urgency at all.

CAUTION: Without urgency, decisions are deferred, infighting is allowed to continue, and priorities are not properly set to make the move successful.

2. Ignoring the schedule compression effect

The closer you get to your move date, the more the schedule appears to accelerate toward your milestones. What's actually happening? The schedule is compressing. The closer you get, the fewer hours in the day you seem to have while needed resources are scarce. The end date doesn't move, yet more work is crammed into the remaining time in one last desperate attempt to meet the milestone.

CAUTION: When reviewing your schedule, pay particular attention to those final months with an eye toward building contingency buffers and resource alternatives. This will help combat schedule compression.

3. Underestimating coordination issues

While the technical challenges are not small, neither is the coordination required across your organization. Coordination is needed with your customers, your vendors, your internal staff, your executive team, and your end users. You wouldn't be the first organization that failed to coordinate with the finance staff and missed payroll, vendor payments, and customer invoices due to an ill-timed move. Integrate all the important business intersections with your move timeline.

CAUTION: Professionals are often overwhelmed during this time. Tailor your coordination to their situation.

4. Failing to contractually obligate vendors

Simply, if you need a vendor to perform a move-related service (such as moving your Storage Area Network), don't leave that to a handshake. Get a statement of work (SOW) that outlines the schedule and deliverables so your key vendors are contractually on board and committed to your requirements.

5. Failing to focus on equally critical post-move issues

These can be different for each organization, but post-move employee retention and de-commissioning of unneeded services are two that rise to the top of most post-relocation plans.

CAUTION: Unexpected staff losses usually occur after a difficult move.

6. Constructing a plan in which EVERYTHING has to go right

Virtually all move project plans suffer from this blunder. Review your plan with others who have deep experience to ensure that key risk areas and contingencies are addressed.

My hope is that this guide clarifies your data center move for you, your staff, and your vendors. Use it to develop your onboarding strategy and jumpstart your plan.

Are you struggling with something specific? Let me know how I can help.

SELF-ASSESSMENT

WHAT ARE THE SIGNS YOU NEED OUTSIDE HELP?

It's not unusual for a data center move to stretch a firm's resources. Experienced outside executive leadership may be needed for a temporary engagement either pre-move to jumpstart governance and move planning or post-move to continue the transformation triggered by the relocation.

What are the signs you need outside help? Some easy-to-spot behaviors include:

- Your procurement department is struggling to write an RFP (Request for Proposal) or Statement of Work (SOW).

- Your project manager is using boilerplate found online.

- Members of your staff have never moved a data center and they are piecing together a recipe from the online bazaar.

- Your schedule is tight and you need your resources focused on their "real" jobs.

- Your inherited schedule and constraints already seem impossible to meet.

Outside help frees your scarce resources, alerts you to hazards, and jumpstarts your efforts.

The value of planning and precise execution of your data center relocation is easy to see. Done well, customers and staff stay informed, downtime is minimized, and your company avoids negative press.

Finding a firm that can deliver what you need when you're unsure of your needs is challenging. Keep in mind that more than one firm might be needed. These suggestions can help you articulate a coherent *needs briefing* to potential partners:

- Understand the data center moving process first; this guide gets you started.

- Make initial decisions about the destination, key milestones, and expected downtime before contacting vendors.

- Understand the urgency. Don't spend all your time on the RFP (Request for Proposal) process and leave little time to plan and accomplish the move. This is like having the shot clock expire while you failed to get off the shot. It's a turnover. Note: Sometimes an RFP is unnecessary and doesn't produce the expected competitive cost-savings anticipated. You could shortlist your vendors first instead.

- Is vendor independence important? Then select one that doesn't sell you hardware and software and will also oversee multiple vendors on your behalf.

- On the other hand, is using a single vendor key to your happiness? Then select a vendor that can supply equipment and services, one that already has deep skill in your technologies. This isn't always the OEM (Original Equipment Manufacturer), but contacting the OEM is a natural starting point.

Select a relocation firm that has a successful track record and can adapt to inevitable changes to your data center moving plan. Vendors who bring an inflexible, one-size-fits-all approach may hamper your progress. That's why it's important to understand their move methodology. Inspect their case studies, contact their references, and ask these questions:

- Do you want the staff that actually performed a vendor's referenced data center moves? Recognize that unless you insist, the people assigned to your move may not be the ones who performed on the references.

- Do you want a turnkey, hands-off move or a collaborative move that involves your key staff members?

- Ask the same questions of the vendor's technical team that you asked of their marketing team. Inconsistencies in their answers should be a red flag requiring additional clarifications.

If cost is your only evaluation criteria, don't drag out the selection process. Avoid paying inexperienced vendors time to learn on your dime. Deep experience can adapt quickly to hazards where inexperience collides with obstacles quite spectacularly.

Be realistic about the pricing model you want. Most contracts are not fixed price because the work scope is rarely fixed. Even with fixed-price contracts, change orders may be required to handle the scope creep. And the approval process of these change orders can drain precious time off the clock—time needed to meet your deadlines.

Most complex data center moves need some type of outside help—from a simple move plan review to full-blown project. Formulate a strategy right from the beginning to select the right team. Your *move narrative* will help you align what vendors provide with what you actually need.

If you're performing the move with all internal resources instead, these tips can help you start right:

- Establish a common sense of urgency.
- Establish the specific governance for this project.
- Over-communicate to all stakeholders.
- Establish a methodology.
- Don't forget to rehearse all moves.

- Budget for staff recognition.
- Be realistic about asking staff to do two jobs simultaneously.

Don't underestimate the amount of time that coordination demands. Above all, remember that a data center move is much more than a technical project. Your business depends on data center functions and you must approach the project from that business perspective.

Find qualified vendors for various aspects of your data center move from the Services Directory at: http://datacentermoving.com/services.html

ABOUT THE AUTHOR

Since founding E-Oasis in 1996, Blaine Berger has moved data centers and provided interim executive leadership to amazing clients. Blaine's previous 13-year career at IBM established a foundation for his deep experience across many industries.

An inventor with several U.S. patents, Blaine is an electrical engineer with a Bachelor of Science degree from the University of Wyoming. He is also certified in Geographical Information Systems from the University of Denver.

You can connect with Blaine on LinkedIn, Google+, or via e-mail blaine@e-oasis.com.

Author's Note

This guide has benefited from feedback from generous readers. Did you find it useful? Are there topics you'd like to see included? Do you have a data center move success story to share? I welcome all of your feedback, suggestions, and questions.

If this guide helped you, please leave an online review.

Blaine Berger

blaine@e-oasis.com

PROMOTIONAL CONSIDERATION

Writing, publishing, and distributing require revisions, updates, and notification and the resources to complete these tasks. Nothing in the relationship between sponsor and author affects the editorial independence of this guide's content.

I welcome the opportunity to provide this benefit to sponsors who wish to amplify their own marketing activities. ***Effective marketing with books can provide a high return on your investment.*** If you are part of the data center ecosystem, this book fits perfectly into your marketing and promotion strategies.

Individual or bulk orders of print copies of this guide are available. For information about sponsoring future or special editions for your trade shows, seminars, and workshops, contact: blaine@e-oasis.com.

I welcome your inquiries about appearances, cross-promotions, and custom marketing ideas.

23999022R00125

Made in the USA
San Bernardino, CA
08 September 2015